ENVIRONMENTAL POLICY FOR 21ST CENTURY

Carlos Andres Garcia Sanchez

Delve
PUBLISHING
www.delvepublishing.com

Environmental Policy for 21st Century

Carlos Andres Garcia Sanchez

Delve Publishing
4164 Lakeshore Road
Burlington, ON L7L 1A4
Canada
www.delvepublishing.com

Email: orders@arclereducation.com

ISBN: 978-1-77956-377-4 (Hardcover)

ABOUT THE AUTHOR

Carlos Andrés García is an environmental specialist with over 14 years of experience in climate change, sustainability, and international cooperation. He has led and managed projects aimed at fostering resilience, poverty reduction, and environmental sustainability. Currently serving as a Climate Change Consultant, Carlos has designed and supervised numerous climate change adaptation projects, securing international funding from bodies such as the Nordic Fund, USAID, and the Green Climate Fund. Throughout his career, Carlos has written numerous scientific articles focused on conservation programs, climate development, and environmental governance. His publications have contributed significantly to advancing knowledge in areas such as climate resilience, marine conservation, and environmental governance. Carlos's professional journey includes his work as a senior consultant for international development banks, where he assessed climate-resilient solutions for sectors like ports, coastal infrastructure, and agriculture. His expertise extends to navigating international environmental standards, promoting Nature-Based Solutions (NbS), and managing large-scale conservation programs. He has played a pivotal role in enhancing the environmental governance of protected areas and supporting local communities in adaptation strategies. In addition to his work in project management, Carlos holds a Master's in Environmental Economics from the University of Western Brittany in France and a Bachelor's in Business Administration from Colombia. He is also a certified Climate Change Officer. His dedication to sustainable development, along with his extensive writing and research experience, makes him a key contributor to climate action efforts across Latin America and globally.

TABLE OF CONTENTS

LIST OF FIGURES

LIST OF TABLES

PREFACE

This book, "Environmental Policy for the 21st Century," provides a comprehensive examination of the current landscape of environmental policy, exploring innovative approaches and critical issues that shape the development and implementation of effective environmental strategies.

It has eight chapters. The first chapter, "Introduction to Environmental Policy in the 21st Century," sets the stage by outlining the key principles and historical context of environmental policy. It provides a foundational understanding of the evolution of environmental governance and its importance in contemporary society. The second chapter delves into "Climate Change Policy," analyzing the global efforts to combat climate change, the role of international agreements, and the effectiveness of various policy instruments. The third chapter, "Area-Based Conservation in the 21st Century," explores strategies for protecting natural habitats and biodiversity through targeted conservation efforts.

Chapters four and five shift focus to renewable energy, with "The Impact of Environmental Policy on Renewable Energy Innovation" examining how policies can drive technological advancements and promote the adoption of renewable energy sources. "Evaluating Renewable Energy Policies" provides a critical assessment of existing policies, identifying successes and areas for improvement. Chapter six, "Environmental Policy (EP) and Preventive Environmental Management (PEM)," discusses the integration of preventive measures in policy-making to prevent environmental degradation. Chapter seven, "Equity, Environmental Justice, and Urban Climate Change," addresses the intersection of social equity and environmental policy, emphasizing the need for inclusive approaches that consider vulnerable communities. Finally, the book concludes with "International Environmental Law," offering insights into the global legal framework for environmental protection and its implementation challenges and opportunities.

This book aims to equip readers with the knowledge and tools to engage with and influence environmental policy meaningfully.

—Author

Chapter **1**

INTRODUCTION TO ENVIRONMENTAL POLICY IN 21ST CENTURY

Contents

INTRODUCTION

Environmental policy focuses on governing the interaction between the natural environment and humans to ensure a mutually beneficial relationship. In the past, environmental policy has been primarily focused on addressing issues such as pollution control, waste management, and habitat preservation. Nevertheless, policymakers increasingly recognize the importance of successfully integrating environmental policy with other policy areas and sectors. From a political perspective, the concept of 'environmental policy integration' signifies a challenging yet crucial phase in the ongoing transformation of environmental policy towards a more comprehensive approach that promotes sustainable human development (Jordan et al., 2013).

Environmental policy in the 21st century represents a critical intersection of scientific understanding, economic necessity, and social responsibility. As we move further into this century, the environmental challenges we face have become more complex. Climate change, biodiversity loss, pollution, and resource depletion are no longer distant threats but immediate realities impacting communities and ecosystems globally. The need for robust, forward-thinking environmental policies has never been more pressing. These policies must address not only the mitigation of environmental damage but also the adaptation to changes that are already underway (Gaines, 2014).

In the past, environmental policy development has been shaped by the need to address various urgent environmental issues. In the late 20th century, notable advancements were made by implementing important international agreements and legislative frameworks focused on safeguarding natural resources and mitigating pollution. Nevertheless, the current measures are falling short of addressing the rapidly worsening environmental degradation witnessed in recent decades. In today's fast-changing world, adopting an integrated and proactive approach to environmental governance is vital. It is crucial to involve the government and private sections, civil society, and private sectors in promoting sustainable practices (Stoeglehner, 2020).

Modern environmental policy is strongly rooted in data and scientific research. With the rapid progress of data collection and technology, humans have an unparalleled understanding of environmental systems and how their activities affect them. Today, policies are being shaped by comprehensive modeling and environmental assessments, allowing for improved prediction and environmental risk management. For instance, climate policies now rely on the scientific assessments of the Intergovernmental Panel on Climate Change (IPCC), which offers rigorous findings on climate change (Barnosky et al., 2014).

Considering economic factors is crucial when shaping environmental policies. The transition to a sustainable economy requires significant investments in green technologies and infrastructure. Policies that promote renewable energy, energy efficiency, and sustainable agriculture are not only environmentally beneficial but also economically advantageous. They have the potential to generate employment opportunities, drive forward new ideas, and result in significant cost reductions by mitigating environmental harm and conserving resources (Cairns, 1993).

1.1. GENERAL CHARACTERISTICS AND HISTORY OF ENVIRONMENTAL POLICY

1.1.1. History

People have been contaminating and depleting natural resources for as long as they have altered their surroundings. Throughout history, there has been a clear connection between economic demands, population growth, and environmental impact. One notable example of this is Europe's significant decline in forests since the twelfth century. This decline was driven by the growing need for timber to meet the demands of a rapidly increasing population. In the 18th century, a response was formulated to address the destructive consequences of over-exploiting forests. This response aimed to ensure a consistent timber supply while preserving the other important functions of the forest (Sutter, 2013).

International environmental policy went through various stages throughout the 20th century, mirroring the condition of the environment, shifts in political awareness of environmental issues, advancements in scientific understanding, the implementation of new technologies, and the evolving framework of international law. Throughout the 1930s to the 1960s, literature emerged rooted in scientific knowledge and played a crucial role in shaping political efforts for environmental protection (Lafferty and Hovden, 2003). In 1962, Rachel Carson released Silent Spring, a powerful critique of pesticides' widespread usage and a plea for a shift in humanity's perception of the natural world. The book quickly captured the interest of many and played a role in increasing public consciousness about the environment in danger. Carson's studies concentrated on the indiscriminate use of insecticides such as DDT. After her article, most nations around the globe, including the United States, outlawed DDT as an insecticide. Today, the book is widely recognized as a significant contribution to the history of environmentalism, with a lasting impact on contemporary thought that extends well beyond the issue of DDT (White, 1985). In 1970, a significant shift in understanding the relationship between humans and nature led the US

Congress to legislate the National Environmental Policy Act (NEPA) and the founding of the Environmental Protection Agency (EPA). It marked a crucial step in addressing the pressing environmental issues of the time. During the late 1960s and early 1970s, there was a growing awareness of environmental issues in industrialized countries. Riots and public demonstrations were occurring, highlighting the urgent need for action. One of the major concerns was environmental pollution, particularly the alarming occurrence of acid rain in northern Europe. During a growing skepticism and public unrest, a noticeable shift away from the status quo occurred. The increasing demand for the preservation of nature resulted in the emergence of green parties in Europe and the convening of the UN Conference on the Human Environment in 1972, also known as the Stockholm Conference (Robin and Griffiths, 2004). It marked the start of a proactive era in international environmental policy, bringing fresh perspectives on the connection between the Earth's natural resources and human society. The Stockholm Conference had a significant impact on regional and global activities and the development of laws and institutions. In 1973, the United Nations Environmental Program (UNEP) was established as the primary environmental hub within the United Nations. Over time, other international organizations began developing their environmental protection initiatives. It included organizations such as the World Bank, United Nations Educational, Scientific and Cultural Organization (UNESCO), World Meteorological Organization (WMO), World Health Organization (WHO), and the Food and Agriculture Organization of the United States (FAO) (Cronon, 1993).

1.1.2. Features of Modern Environmental Policy

Since the term "environment" first appeared in the early 1960s, our understanding of how humans affect the natural world has expanded quickly. Over time, there was a growing understanding of the impact and emission of various sources of pollution, and fresh kinds of pollution were gradually discovered. Ecology as a science is now taken seriously by the public and most scientists. In the 1970s, there was a significant rise in ecological studies and the development of new technologies to study the environment, including radar for observing the Earth's surface. During the 1980s, scientific matters started to take on a political dimension (Glasbergen, 1998). Global and regional environmental threats gained significant attention on the international community agenda as scientific evidence highlighted the possible consequences of biodiversity loss, ozone depletion, and climate change. Local problems were acknowledged to have global or transboundary effects on humanity, such as desertification or the disappearance of tropical forests. The political nature of environmental policy formulation is often defined by the delay between cause and effect, even while science dominates the substance of the process (Jordan and Lenschow, 2010). Given the pressing environmental concerns, such as the alarming deforestation rates, rising greenhouse gas (GHG) emissions, and declining biodiversity, it is evident that effective long-term strategies are needed in environmental policy. Time is a crucial factor in national and international environmental policy. The experiences of the last few decades of the 20th century show that democratic regimes are more suited than totalitarian or authoritarian ones to handle long-term problems. However, it is important to consider that political democracies and market economies often prioritize short-term concerns, such as shareholder value or voting results, which may limit their consideration of long-term aspects (Jendrośka, 2021). Regarding practical matters, party leaders must consider balancing achieving immediate wins and implementing sustainable long-term policies. It is important to consider that the environmental impacts of social or economic policies may not directly impact the generations or individuals who benefit from them. This lack of immediate consequences has led to the need for sustainable development and intergenerational equity. Furthermore, the fact that science is not always certain about the cause and effect of environmental degradation necessitates legislative action to tackle these concerns (Jaffe et al., 2002).

First introduced in the 18th century by German forest managers, sustainable development gained prominence in the 'Brundtland Report' of 1987. This report emphasized the importance of development in fulfilling present needs while ensuring that future generations can do the same. Understanding sustainable development can be challenging as it encompasses multiple policy concerns. It emphasizes the need for policy-makers to consider environmental, economic, and social factors while ensuring an equal balance between the needs of future and present generations. Sustainable development is now widely acknowledged as the central idea in international environmental and development policy, having made multiple appearances in international agreements and other instruments. Given each State's wide range of cultural, economic, and social conditions, it is challenging to develop a single operational-oriented definition for this concept. At the global level, one of the main challenges in environmental policy is finding a practical and efficient consensus on the definition of sustainable development that considers the circumstances of both developing and industrialized countries (Chen et al., 2021).

1.2. ENVIRONMENTAL POLICY AREAS

Environmental problems can be categorized based on their origin and impact, including global, international, regional, and local issues. Recognizing that pollution and degradation stem from a single local point is important. Still, it is also valuable to consider and develop suitable policy responses at the global, regional, and local levels (Knill et al., 2010).

The various concerns and environmental policy areas are incredibly diverse and complex.

a. Global climate change is a significant environmental challenge. It appears that the rise in greenhouse gas levels in the troposphere and the artificial warming of the Earth's atmosphere are primarily caused by human activities, such as using fossil fuels, changes in land use, and agricultural practices. Scientific consensus warns of a concerning rise in temperature, projected to be between 1.4°C and 5.8°C by 2100 compared to 1990. Moreover, by 2100, the average sea level is expected to rise by 9 cm to 88 cm. These changes would significantly affect food production, public health, the environment, and economies (Piorr, 2003).

b. The rapid deterioration of numerous forests worldwide poses a significant challenge for the international community. Numerous factors contribute to the significant destruction of forests today, particularly in tropical regions. These include the increasing human needs of rising populations in developing nations, diseases, pests, overgrazing, air pollution, forest fires, unsustainable commercial logging, industrial wood and fuelwood, industrial wood overlogging, and agricultural expansion (Arts et al., 2006).

c. Ecosystem destruction and extinction are closely linked to a lesser-known decline in biodiversity. It is widely acknowledged that a significant percentage (50–90%) of the world's species reside in tropical forests. The rapid decline of natural habitats and species globally is a cause for great concern, as it may have severe and unpredictable consequences in the future.

d. Soil degradation is a significant concern that has impacted approximately one-third of the world's agricultural soils, often resulting in desertification. Deserts make up approximately 7 % of the Earth's land surface. The desertification poses a threat to the livelihoods of more than a billion individuals in over 100 nations, leading to a decline in the productivity of grazing land and farming (Gibbs and Jonas, 2000).

e. Over the past decades, addressing water and air pollution has seen some success in industrialized states. However, this issue worsens in many parts of the developing world, especially in the ever-expanding megacities.

f. One of the key factors driving changes in land use is the rapid growth of the economy and the increasing need for food to sustain a growing population. It is projected to be a significant rise in global food demand, with a staggering 110 % increase by 2050. It will undoubtedly put additional strain on the last non-domesticated areas across the globe (Fiorino, 2001).

g. High population growth, improving living standards, and ongoing industrialization contribute to an increased demand for fossil energy. The energy use per capita in industrialized countries is already at an alarmingly high level and shows no signs of slowing down. Meanwhile, smaller developing nations were projected to significantly increase their share of commercial energy consumption, reaching nearly 40 % by 2010. Developed and developing nations are likely to see increased energy use in the transportation industry as more people drive private vehicles due to increased earnings. Energy consumption contributes to both global and local air pollution issues (Russi et al., 2016).

h. Ecological damage poses a threat to oceans, not only to the Earth's land surface of Earth. International waters face significant challenges due to marine resource exploitation and water quality degradation. Human activities on land are responsible for significant marine pollution, particularly in coastal areas. These zones are crucial for fishing, accounting for nearly 90 % of the world's catch, and are also home to 75% of the global population (Hertin and Berkhout, 2003).

i. Another major issue on the international and national environmental agenda is the transportation and disposal of hazardous waste from industry and household sources. Many chemical wastes and industrial goods are either difficult to dispose of or nondegradable. With the rise of modern tourism, new tourism destinations have significantly expanded, especially in the Caribbean, Latin America, North Africa, and Asia. Unfortunately, this growth has not only brought economic benefits but also led to an alarming degradation of the environment in these regions (Hahn, 2000).

1.3. INSTRUMENTS OF ENVIRONMENTAL POLICY

Environmental policy interventions utilize various instruments to address failures and regulate behavior to achieve society's environmental goals (González, 2011).

1.3.1. Constitutional Provisions

Constitutional provisions that safeguard the environment serve as a crucial component of national environmental law. One notable example is Art. 20a of the Basic Law of the Federal Republic of Germany, which was implemented in 1994. According to this Article, the State must safeguard the environment for the well-being of present and future generations. This responsibility is to be fulfilled through legislation and the exercise of judiciary and executive powers. This rule recognizes the importance of environmental protection as a state goal; however, it does not directly grant any rights to citizens (Keith, 2002).

1.3.2. Command-and-Control Approaches

Two main types of tools can be used for environmental policy. On the one hand, there are reactive command-and-control approaches (CAC), which focus more on responding to environmental issues. On the other hand, there are preventive market-based instruments that aim to create economic incentives for individuals and individuals to minimize environmental harm. Throughout the 1970s and 1980s, governments relied on CAC regulations as a common approach. After economic studies revealed that CAC regulations are frequently expensive, rigid, and ineffective, this strategy was reexamined. When the government imposes flat CAC rules, the private sector often lacks the motivation to invest in creating new and improved control technologies or strategies (Gomes et al., 2024).

1.3.3. Market-Based Instruments

Due to the shortcomings of CAC regulations and the importance of encouraging polluters to prioritize environmental innovation, alternative instruments have been developed to better align environmental and economic decision-making within the legal and political framework. Market-based policies give markets and private industry the freedom to choose the most efficient methods of achieving environmental goals. They also promote cooperation and flexibility between industry and government. However, it is important to note that market-based instruments still require government supervision. A wide range of market-based policies are currently in place (Pirard and Lapeyre, 2014).

Since its introduction in the USA in 1968, emissions trading has undergone significant changes in the perceptions and attitudes of policy-makers. This innovative concept enables polluters to benefit from the efforts of more effective actors in the same industry by compensating them for their additional efforts. There has been a significant increase in interest in emissions trading, particularly in controlling GHG gases. This growth can be attributed to the successful implementation of a tradable permit system in the USA, which effectively regulated ozone-depleting substances and sulfur dioxide. Article 17 of the (Langniss and Praetorius, 2006), an addendum to the United Nations Convention on Climate Change, envisions a unique greenhouse gas trading scheme. Emissions trading is a straightforward concept. Every participant must adhere to a designated emission limit and has the option to trade portions of the emission allowances with other entities. Regarding the economy, emitters with fewer abatement costs can choose to reduce their emissions and sell any extra allowances they have (Gómez and Muradian, 2015).

In comparison, polluters with higher abatement costs will likely opt to purchase these allowances instead. It is possible to decrease the total allowable emissions over time gradually. Two additional types of flexible market-based innovations are allowed under the Kyoto Protocol. These include projects in developing states and Joint Implementation among industrialized countries governed by the Clean Development Mechanism (Henderson and Norris, 2008).

One way to encourage pollution prevention is through industry self-commitments, such as European car manufacturers voluntarily reducing fuel consumption. Voluntary programs can be implemented more quickly than regulations, resulting in faster reductions. Because volunteer programs are still relatively new, it might be challenging to quantify their effectiveness (Klassert and Möckel, 2013).

Positive market incentives can be created through various means, such as tax allowances, soft loans, grants, or special environmental taxes (called 'eco-taxes') imposed on products that harm the environment during production, consumption, or disposal. Unfortunately, subsidies granted for reasons other than environmental concerns can lead to environmental degradation. It is important to consider the potential negative impact on the environment of certain policies, such as tax preferences for private vehicle transport, tax preferences for mineral and oil extraction, agricultural commodity price programs, and below-market financing costs for electricity infrastructure. These measures may inadvertently encourage the emission of harmful pollutants, the excessive use of natural resources, and the generation of waste. A debate has been initiated to eliminate or reduce these subsidies, but it has not yet significantly impacted state practice. Public discussions regarding environmental reform are frequently dominated, very appropriately, by the financial implications of environmental policies for the public coffers. One significant way to address these fiscal issues is by removing subsidies (Mazaheri et al., 2022).

The environmental impact assessment is widely recognized as a fundamental tool in today's environmental policy. It establishes a comprehensive evaluation process for various private or public projects, such as motorways, airports, power stations, and even certain infrastructural or agricultural projects. These assessments aim to offer governmental authorities and private entrepreneurs enhanced environmental information regarding a project's potential impacts during the planning phase (Boisvert et al., 2013).

1.3.4. Environmental Policy and Neoliberalism

While the economic neoliberal paradigm offers optimism for increased wealth, it also raises concerns about environmental preservation. The argument for a lean state facilitates and supports a liberal economy, which has dominated discussions at the international and national levels during the 1990s and has been strengthened by economic globalization. Economic neoliberalism must be paired with intergovernmental or governmental structures qualified and equipped to obtain enough data and exercise enough authority to safeguard the environment as a public good since free markets cannot adequately price adverse environmental effects. This applies, in principle, to both the international and national level (Arsel and Büscher, 2012).

1.4. PRINCIPLES OF INTERNATIONAL ENVIRONMENTAL RELATIONS

Furthermore, a wide range of principles, rules, and political goals are currently being developed to address the international impacts of environmental issues. These efforts complement the existing legislation that outlines the rights and obligations of those involved in international or national environmental law (Ong, 2010).

1.4.1. Natural Resources Sovereignty

Environmental issues transcend national boundaries. Therefore, the emergence of environmental issues presents fresh obstacles to state sovereignty. This concept has traditionally emphasized the autonomy of states in making decisions about their affairs, including formulating policies regarding the natural environment inside their borders. Thus, sovereignty over natural resources grants a state the authority to safeguard and, unfortunately, even harm or potentially devastate the environment within its borders (Gess, 1964). Nevertheless, there is a limit to this right when environmental harm extends beyond national borders and actions within one State's control negatively impact another state. It is widely acknowledged in international law that the responsibility to safeguard the rights of other nations within their borders is a fundamental principle. This principle is often referenced in international agreements and declarations concerning environmental protection, such as the Stockholm Declaration (principle 21) (Saul, 2013).

Given the current environmental challenges, such as loss of biodiversity and forests, eroding soils, polluted waters, global warming, and ozone depletion, states must take responsibility for both the well-being of their territory and the overall health of the Earth's ecosystem. Therefore, states develop sector-specific regimes that acknowledge the importance of collectively addressing regional and global environmental management to benefit their national interests (Wälde, 1983).

1.4.2. Common but Differentiated Responsibility

Recognizing the importance of collective action and responsibility in environmental protection, it is essential to acknowledge that developing countries and detribalized nations have varying capabilities and responsibilities. This principle of common but differentiated responsibility highlights this distinction. Numerous issues in the realm of environmental protection have emerged due to industrial production in developed countries. Due to the earlier stages of development in developing countries, they now prioritize their own social and economic progress, even if it means polluting the environment as industrialized nations did in the past. Therefore, including developing states in global environmental regimes continues to be a crucial aspect of developing international environmental policies (Stone, 2004).

Considering the potential conflict between ecological responsibility and economic necessity, international treaties on environmental issues often include provisions that acknowledge the unique circumstances and specific needs of developing countries. Specifically, these rules usually focus on funding and technology transfer. Currently, the principle of common but differentiated responsibility is a political concept that reflects options and practical conditions but is not legally binding (Castro, 2016).

1.4.3. Scientific Uncertainty and the Precautionary Principle

When it comes to decision-making on international environmental issues, it is crucial to consider the presence of scientific uncertainty. This is because new scientific information can either validate the assessment of risks or reveal that the risk is less serious than initially estimated. A nation must take proactive steps to prevent environmental degradation, even without complete scientific knowledge, per the precautionary principle (Bodansky, 1991). Although initially applied to marine issues, the precautionary principle has been broadened to nonmarine concerns like climate change and ozone depletion. The Convention for the Protection of the Ozone Layer initially mentioned the precautionary principle. It is also embodied in principle 15 of the Rio Declaration. Despite being mentioned in various international declarations, there is still ongoing debate surrounding the basis and scope of the precautionary principle. The increasing practical acceptance of the concept was evident in the effective negotiation of the Biosafety Protocol for the Biodiversity Convention in 2000 (Gollier and Treich, 2003).

1.4.4. Preventive Principle

It is widely recognized that taking preventive action is far more effective in addressing pollution than dealing with its consequences or assigning blame for environmental damage. It is crucial for states to proactively implement measures to prevent environmental damage and protect the environment. Several international treaties, the 1982 World Charter for Nature, the 1972 Stockholm Declaration, and other documents have all supported the preventive approach. Principle 11 of the 1992 Rio Declaration emphasizes the importance of states enacting effective environmental legislation (Nation et al., 2003). The preventive principle entails the ongoing duty to consider environmental concerns in every governmental control and activity area. It is included in environmental and more general treaties like the European Community Treaty (ECT). Utilizing tools such as enhanced monitoring provisions, risk assessments, and early-warning systems demonstrates a proactive approach. It is crucial to guarantee that these principles are effectively implemented in every State, regardless of differences in national legal systems, to benefit the environment (Moulaee and Lotfi, 2019).

1.5. INTERNATIONAL ENVIRONMENTAL LAW

1.5.1. Global Issues

International environmental issues have become a growing concern for national governments, as the degradation of the environment presents significant challenges to maintaining national sovereignty. The reality of the situation shapes international policies addressing environmental degradation. In the long term, willingly accepting limitations on national sovereignty through international agreements may be more beneficial for national freedom than insisting on complete legal sovereignty without binding commitments. Despite its importance, understanding and accepting this paradox in environmental policy is still not widespread (Ferronato and Torretta, 2019) (Figure 1.1).

During the nineteenth century, bilateral fisheries treaties were established to address overexploitation and protect the environment. However, the focus on border-related environmental issues limited their practical impact. In the early stages of developing international environmental rules, there was a strong emphasis on conserving wildlife, seas, and rivers. This led to the adoption of the first whaling convention in 1931. In 1945, the creation of the UN and the specialized agencies marked a significant milestone in developing international environmental policy (Bateman et al., 2001).

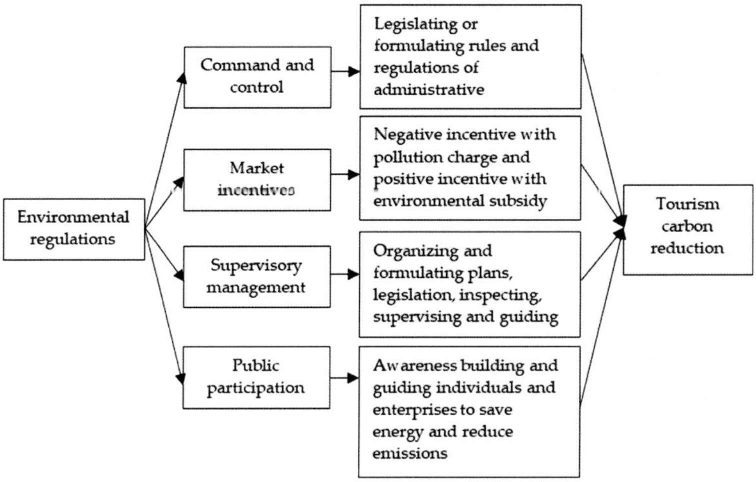

Figure 1.1. *Illustration of the impact of the environment law.*

Source: Chen Q, Mao Y, Morrison AM. (2021). Impacts of environmental regulations on tourism carbon emissions. International Journal of Environmental Research and Public Health, 18(23):12850. https://doi.org/10.3390/ijerph182312850, Creative Commons License.

Over time, international organizations at both global and regional levels started to address environmental issues. These concerns now focus on the pollution caused by specific ultra-hazardous activities. Unfortunately, there was little awareness of the connection between environmental protection and economic development. Despite the increasing focus on environmental protection, significant international, national, and local efforts have been made. The first significant milestone occurred in 1972 at the Stockholm Conference; at this conference, governments aimed to safeguard and improve the quality of the human environment for both current and future generations, as stated in Principle 1 of the Stockholm Declaration on the Human Environment. During the mid-1980s, there was a noticeable shift towards internationalization as states recognized the growing interdependence and the need for new frameworks to address environmental challenges. In 1985, the first treaty for ozone layer protection was concluded, alongside the adoption of new global and regional conventions addressing environmental issues (Widdowson, 1997). In 1992, the United Nations Conference on Environment and Development in Rio de Janeiro (UNCED) marked a significant milestone in international negotiations on universal environmental issues. This conference brought together 176 states and numerous nongovernmental organizations, paving the way for important discussions and actions. UNCED was focused on a fresh approach to environmental politics, specifically the pursuit of a harmonious balance between economic development and environmental protection. AGENDA 21, the Principles on Forests, and The Rio Declaration are three nonbinding instruments adopted by the conference in addition to two legally binding treaties, the Framework Convention on Climate Change and the Convention on Biological Diversity (Grunwald et al., 2011). It is crucial to incorporate the principles of sustainable development and its ecological aspects into economic policies, as emphasized by UNCED. The World Bank, in particular, has emphasized, more than other international organizations, the importance of a comprehensive approach to addressing environmental, social, and economic issues. They have begun developing broad development frameworks with their client states. There is a growing agreement on the significance of technology cooperation in achieving sustainable development. Developing countries emphasize the need for the transfer of technical equipment and machinery. In contrast, developed countries focus on sharing knowledge of science and technology and providing support to enhance institutional capacity in addressing environmental concerns (Braun, 2006).

Following UNCED, several institutional innovations have been implemented, such as establishing the Commission of Sustainable Development (CSD) under the United Nations. In addition, funds have been established to facilitate the transfer of financial resources and technology to specific sectors, such as the Multilateral Ozone Fund (Ramakrishna et al., 2006). The Global Environment Facility (GEF) was established in 1990 and underwent significant reforms in 1994. It plays a crucial role in funding projects related to global environmental protection, with a specific focus on four key areas: protecting the ozone layer, addressing climate change, managing international waters, and preserving biological diversity. Today, the GEF is widely recognized as an institution that promotes environmental cooperation between different regions, ensuring a fair voting structure that prevents one region from dominating. While industrialized states have formally acknowledged in the GEF that developing states must receive support from the developed world for successful global environmental policy to be implemented, developing states have also acknowledged that efforts within their own societies and economies are necessary to address global issues. Indeed, the current funding for the GEF is limited to less than US$1 billion per year (Leong et al., 2021).

In 1997, a significant event occurred when the 19th UN General Assembly Special Session (UNGASS) convened to assess the advancements made in sustainable development, as outlined in the Rio Declaration and Agenda 21 (also known as 'Earth Summit +5'). Five years after Rio, there had been some encouraging outcomes. States demonstrated a growing readiness to embrace international treaties to safeguard the global climate, the ozone layer, and biodiversity and address the soil degradation and expansion of deserts. Various segments of civil society have come together to tackle the extensive range of problems on the environmental agenda. Despite the efforts made by professional speechwriters, it is undeniable that significant environmental problems persist worldwide, and the global environment is still deteriorating. International environmental policy-making faces significant challenges due to the lack of strong and influential organizations like the International Monetary Fund (IMF) and the World Trade Organization (WTO). This institutional deficiency shall grow increasingly apparent as national governments continue to prioritize their short-term concerns and own agendas. Therefore, enhancing institutions will continue to be a priority at the global level to improve policy-making (Shiffman, 2009).

1.5.2. Regional Issues

It is important to consider the environmental regulations of regional organizations within their broader political and legal framework. Therefore, the environmental law and policy of the EC were greatly impacted by the Treaty of Amsterdam (October 2, 1997), which brought about significant alterations to the environmental provisions of the EC treaty (ECT). One of the objectives of the European Community is now officially recognized as sustainable development (Article 2, ECT). In addition, the concept of integration, which calls for incorporating environmental policy considerations into various aspects of community policy, is now included in a single provision at the beginning of the ECT (Article 6, ECT) (Irwin et al., 2010).

Throughout the negotiations of NAFTA, environmental concerns played a significant role as Canada, the US, and Mexico worked towards liberalizing regional trade. Although the agreement included various environmental provisions such as safety, health, and environment standards and measures for phytosanitary and sanitary concerns, the US environmental community perceived these provisions as insufficient to address the environmental risks associated with the increased trade between the three nations. As a seasoned expert in environmental policy, it is worth noting that in response to this criticism, the parties involved engaged in negotiations, creating a Side Agreement known as the North American Agreement on Environmental Cooperation (NAAEC). This agreement came into effect on January 1, 1994, alongside NAFTA (Butler, 1996).

Given the growing significance of regional economic associations, developing environmental policies that are not undermined or disregarded when promoting economic advancement is crucial (Xu, 2008).

1.6. SUMMARY

The current State of environmental policy is at a crucial juncture in its relatively brief existence. During the 1970s, governments worldwide took action in response to the growing importance of environmental concerns. They created agencies and implemented laws to tackle specific issues as they emerged. Policy-makers often viewed environmental issues as primarily technical challenges, resulting in a predominantly reactive and sector-focused approach to policy-making. However, in the 1980s, the conversation regarding environmental issues shifted towards sustainable development. It is important to note that sustainable development goes beyond just economic progress with the addition of waste reduction and cleaner technologies. It is a comprehensive approach that aims to meet environmental, economic, and social objectives, both now and for future generations.

In the modern era, there is a growing consensus that environmental considerations should be incorporated into policy domains that have traditionally been seen as separate from the environment, such as energy, transport, and agriculture. Over the past four decades, it has become evident that the root causes of unsustainable behavior are deeply embedded within these sectors. Without addressing these underlying causes, environmental protection would perpetually be ad-hoc, piecemeal, and reactive. From a political perspective, the shift towards environmental policy integration marks a fresh and potentially challenging phase in the ongoing evolution of environmental policy. Nevertheless, the political implications are significant as environmental policy gradually erodes its independent existence by incorporating green initiatives into other policy areas. Interestingly, when the environment is thoroughly integrated into all aspects of policy-making, the need for a distinct 'environmental' policy becomes obsolete. However, it is still up for debate whether this result will actually happen in reality.

REFERENCES

1. Arsel, M., & Büscher, B. (2012). Nature™ Inc.: Changes and continuities in neoliberal conservation and market-based environmental policy. *Development and Change, 43*(1), 53–78.
2. Arts, B., Leroy, P., & Van Tatenhove, J. (2006). Political modernisation and policy arrangements: A framework for understanding environmental policy change. *Public Organization Review, 6*, 93–106.
3. Barnosky, A. D., Brown, J. H., Daily, G. C., Dirzo, R., Ehrlich, A. H., Ehrlich, P. R., & Wake, M. H. (2014). Introducing the scientific consensus on maintaining humanity's life support systems in the 21st century: Information for policy makers. *The Anthropocene Review, 1*(1), 78–109.
4. Bateman, C., Baker, T., Hoornenborg, E., & Ericsson, U. (2001). Bringing global issues to medical teaching. *The Lancet, 358*(9292), 1539–1542.
5. Bodansky, D. (1991). Law: Scientific uncertainty and the precautionary principle. *Environment: Science and Policy for Sustainable Development, 33*(7), 4–44.
6. Boisvert, V., Méral, P., & Froger, G. (2013). Market-based instruments for ecosystem services: Institutional innovation or renovation? *Society & Natural Resources, 26*(10), 1122–1136.
7. Braun, B. (2006). Environmental issues: Global natures in the space of assemblage. *Progress in Human Geography, 30*(5), 644–654.
8. Butler, R. W. (1996). The development of tourism in frontier regions: Issues and approaches. In *Frontiers in Regional Development* (pp. 213–229).
9. Cairns Jr, J. (1993). Environmental science and resource management in the 21st century: Scientific perspective. *Environmental Toxicology and Chemistry: An International Journal, 12*(8), 1321–1329.
10. Castro, P. (2016). Common but differentiated responsibilities beyond the nation state: How is differential treatment addressed in transnational climate governance initiatives? *Transnational Environmental Law, 5*(2), 379–400.
11. Chen, Q., Yu, H., & Wang, Y. (2021). Research on modern marine environmental governance in China: Subject identification, structural characteristics, and operational mechanisms. *International Journal of Environmental Research and Public Health, 18*(9), 4485.
12. Cronon, W. (1993). The uses of environmental history. *Environmental History Review, 17*(3), 1–22.

13. Ferronato, N., & Torretta, V. (2019). Waste mismanagement in developing countries: A review of global issues. *International Journal of Environmental Research and Public Health, 16*(6), 1060.
14. Fiorino, D. J. (2001). Environmental policy as learning: A new view of an old landscape. *Public Administration Review, 61*(3), 322–334.
15. Gaines, S. E. (2014). Reimagining environmental law for the 21st century. *Environmental Law Reporter News & Analysis, 44*, 10188.
16. Gess, K. N. (1964). Permanent sovereignty over natural resources: An analytical review of the United Nations Declaration and its genesis. *International & Comparative Law Quarterly, 13*(2), 398–449.
17. Gibbs, D., & Jonas, A. E. (2000). Governance and regulation in local environmental policy: The utility of a regime approach. *Geoforum, 31*(3), 299–313.
18. Glasbergen, P. (1998). Modern environmental agreements: A policy instrument becomes a management strategy. *Journal of Environmental Planning and Management, 41*(6), 693–709.
19. Gollier, C., & Treich, N. (2003). Decision-making under scientific uncertainty: The economics of the precautionary principle. *Journal of Risk and Uncertainty, 27*, 77–103.
20. Gomes, J. E. C., Ehlert, R. R., Boesche, R. M., Santos de Lima, V., Stocchero, J. M., Barone, D. A., & de Araujo Fernandes, R. Q. (2024). Surveying emerging network approaches for military command and control systems. *ACM Computing Surveys, 56*(6), 1–38.
21. Gómez-Baggethun, E., & Muradian, R. (2015). In markets we trust? Setting the boundaries of market-based instruments in ecosystem services governance. *Ecological Economics, 117*, 217–224.
22. González-Eguino, M. (2011). The importance of the design of market-based instruments for CO2 mitigation: An AGE analysis for Spain. *Ecological Economics, 70*(12), 2292–2302.
23. Grunwald, S., Thompson, J. A., & Boettinger, J. L. (2011). Digital soil mapping and modeling at continental scales: Finding solutions for global issues. *Soil Science Society of America Journal, 75*(4), 1201–1213.
24. Hahn, R. W. (2000). The impact of economics on environmental policy. *Journal of Environmental Economics and Management, 39*(3), 375–399.
25. Henderson, B., & Norris, K. (2008). Experiences with market-based instruments for environmental management. *Australasian Journal of Environmental Management, 15*(2), 113–120.
26. Hertin, J., & Berkhout, F. (2003). Analysing institutional strategies for

environmental policy integration: The case of EU enterprise policy. *Journal of Environmental Policy & Planning, 5*(1), 39–56.

27. Irwin, E. G., Isserman, A. M., Kilkenny, M., & Partridge, M. D. (2010). A century of research on rural development and regional issues. *American Journal of Agricultural Economics, 92*(2), 522–553.
28. Jaffe, A. B., Newell, R. G., & Stavins, R. N. (2002). Environmental policy and technological change. *Environmental and Resource Economics, 22*, 41–70.
29. Jendrośka, J. (2021). Roots of modern environmental law in Europe. *Journal for European Environmental & Planning Law, 18*(3), 201–224.
30. Jordan, A., & Lenschow, A. (2010). Environmental policy integration: A state of the art review. *Environmental Policy and Governance, 20*(3), 147–158.
31. Jordan, A., Wurzel, R. K., & Zito, A. R. (2013). Still the century of 'new' environmental policy instruments? Exploring patterns of innovation and continuity. *Environmental Politics, 22*(1), 155–173.
32. Keith, L. C. (2002). Constitutional provisions for individual human rights (1977–1996): Are they more than mere "window dressing?." *Political Research Quarterly, 55*(1), 111–143.
33. Klassert, C., & Möckel, S. (2013). Improving the policy mix: The scope for market-based instruments in EU biodiversity policy. *Environmental Policy and Governance, 23*(5), 311–322.
34. Knill, C., Debus, M., & Heichel, S. (2010). Do parties matter in internationalised policy areas? The impact of political parties on environmental policy outputs in 18 OECD countries, 1970–2000. *European Journal of Political Research, 49*(3), 301–336.
35. Lafferty, W., & Hovden, E. (2003). Environmental policy integration: Towards an analytical framework. *Environmental Politics, 12*(3), 1–22.
36. Langniss, O., & Praetorius, B. (2006). How much market do market-based instruments create? An analysis for the case of "white" certificates. *Energy Policy, 34*(2), 200–211.
37. Leong, H. Y., Chang, C. K., Khoo, K. S., Chew, K. W., Chia, S. R., Lim, J. W., & Show, P. L. (2021). Waste biorefinery towards a sustainable circular bioeconomy: A solution to global issues. *Biotechnology for Biofuels, 14*, 1–15.
38. Mazaheri, M., Roca, J. B., Markus, A., & Walrave, B. (2022). Market-based instruments and sustainable innovation: A systematic literature review and critique. *Journal of Cleaner Production, 373*, 133947.

39. Moulaee, A., & Lotfi, H. (2019). The fragility of the borders of precautionary and preventive principles. *International Law Review, 36*(60), 269–290.
40. Nation, M., Crusto, C., Wandersman, A., Kumpfer, K. L., Seybolt, D., Morrissey-Kane, E., & Davino, K. (2003). What works in prevention: Principles of effective prevention programs. *American Psychologist, 58*(6, 7), 449.
41. Ong, D. (2010). From 'international' to 'transnational' environmental law? A legal assessment of the contribution of the 'Equator Principles' to international environmental law. *Nordic Journal of International Law, 79*(1), 35–74.
42. Piorr, H. P. (2003). Environmental policy, agri-environmental indicators, and landscape indicators. *Agriculture, Ecosystems & Environment, 98*(1–3), 17–33.
43. Pirard, R., & Lapeyre, R. (2014). Classifying market-based instruments for ecosystem services: A guide to the literature jungle. *Ecosystem Services, 9*, 106–114.
44. Ramakrishna, S., Fujihara, K., Teo, W. E., Yong, T., Ma, Z., & Ramaseshan, R. (2006). Electrospun nanofibers: Solving global issues. *Materials Today, 9*(3), 40–50.
45. Robin, L., & Griffiths, T. (2004). Environmental history in Australasia. *Environment and History, 10*(4), 439–474.
46. Russi, D., Margue, H., Oppermann, R., & Keenleyside, C. (2016). Result-based agri-environment measures: Market-based instruments, incentives or rewards? The case of Baden-Württemberg. *Land Use Policy, 54*, 69–77.
47. Saul, B. (2013). China, natural resources, sovereignty, and international law. *Asian Studies Review, 37*(2), 196–214.
48. Shiffman, J. (2009). A social explanation for the rise and fall of global health issues. *Bulletin of the World Health Organization, 87*(8), 608–613.
49. Stoeglehner, G. (2020). Strategicness–The core issue of environmental planning and assessment of the 21st century. *Impact Assessment and Project Appraisal, 38*(2), 141–145.
50. Stone, C. D. (2004). Common but differentiated responsibilities in international law. *American Journal of International Law, 98*(2), 276–301.
51. Sutter, P. S. (2013). The world with us: The state of American environmental history. *The Journal of American History, 100*(1), 94–119.
52. Wälde, T. (1983). Permanent sovereignty over natural resources: Recent developments in the mineral sector. In *Natural Resources Forum* (Vol. 7, No. 3, pp. 239–251). Oxford, UK: Blackwell Publishing Ltd.

53. White, R. (1985). American environmental history: The development of a new historical field. *Pacific Historical Review, 54*(3), 297–335.
54. Widdowson, H. G. (1997). EIL, ESL, EFL: Global issues and local interests. *World Englishes, 16*(1), 135–146.
55. Xu, J. (2008). Governing city-regions in China: Theoretical issues and perspectives for regional strategic planning. *Town Planning Review, 79*(2, 3), 157–186.

Chapter 2

CLIMATE CHANGE POLICY

Contents

INTRODUCTION

Extensive research in the geophysical sciences has shown that climatic changes, such as long-term shifts in precipitation and temperature patterns, have been a regular occurrence throughout Earth's history. Furthermore, the impact of human activities on climatic conditions has significantly increased since the Industrial Revolution in the late 18th century. Compositional changes to the Earth's atmosphere play a significant role in driving climatic changes. There has been a significant rise in GHG emissions, particularly methane and carbon dioxide, resulting from the combustion of fossil fuels and activities in forestry and agriculture (Gowdy, 2008). Consequently, the atmosphere now contains greater concentrations of these gases. Increased concentrations of greenhouse gases lead to elevated surface temperatures on Earth. This is due to the trapping of extra energy within the atmosphere. The consequences of higher temperatures are numerous and detrimental to both humans and nature. They include intensified rising sea levels, extreme weather events, floods, and droughts (Pindyck, 2013).

Given the significant impact of climate change on both human and non-human life, it has become a subject of immense interest for researchers in both natural and social sciences. Political scientists have given relatively little attention to the issue of climate change in contrast to other social issues such as wars, elections, and the welfare state (Weber and Peters, 2009). There has been a noticeable disparity in the coverage of climate politics between overall political science journals and the prominent natural sciences journals. Experts in environmental policy have primarily published their research in specialized journals of the discipline, such as Global Environmental Politics. They have also contributed to inter- and multidisciplinary journals like the Proceedings of the National Academy of Sciences, Nature Climate Change, Ecological Economics, Climatic Change, and Global Environmental Change (Grubb, 2004).

Given the increasing recognition of climate politics in prevalent political science research, it is a good time to assess this field's current state of research. This is an attempt to highlight the advancements that have been achieved and the areas that still require attention. Additionally, it provides insights into potential research directions in the coming years (Gupta, 2010) (Figure 2.1).

Efforts to address the global climate change issue through political negotiations and global treaties are moving much slower than most climate scientists consider necessary to prevent significant climate changes. Beginning with an analysis of research considering the challenges associated with attaining global climate cooperation (Ripple et al., 2014). While the causes of these challenges are now well understood, there is still much to learn about three other important issues that are crucial scientifically and practically. One important aspect to consider is the design of institutions, which can play a crucial role in addressing and resolving key challenges in global cooperation. The second focuses on factors that influence local and national climate policy differences (Schmidt et al., 2008). The third focuses on factors influencing climate policy outside government, specifically civil society, the relationship between science and policy, and public opinion. After thoroughly examining the literature on these issues, it is not wrong to conclude that significant and inevitable climatic changes will occur if significant efforts to reduce GHG emissions are not initiated shortly. Political scientists have also examined the potential sociopolitical impacts of climate change (Scott and Becken, 2010).

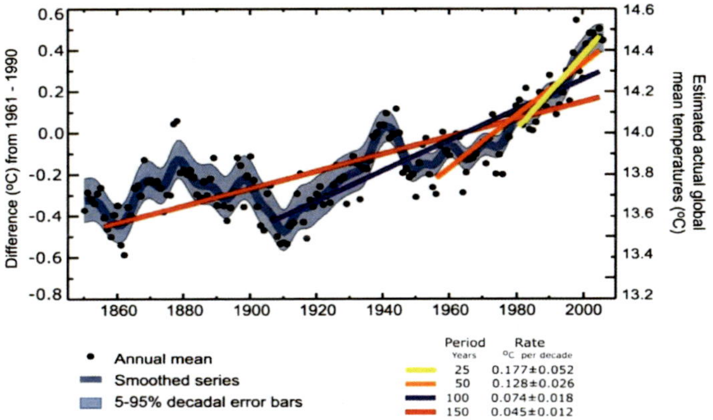

***Figure 2.1.** The rate of increase in temperature is estimated for several different time intervals.*

Source:** Akasofu, S. I. (2013). On the present halting of global warming. Climate, 1, 4–11.* https://doi.org/10.3390/cli1010004).***

Given the gaps highlighted in this effort, the final section will present potential avenues for future political science research. These recommendations are arranged in order from micro to macro. Given the slow progress of global governance efforts, it is worth shifting our perspective to focus more on bottom-up dynamics for normative and analytical reasons. This change in approach is motivated by the need to address the current challenges effectively (Shove, 2010).

2.1. WHY IS IT DIFFICULT TO SOLVE GLOBAL PROBLEMS?

Experts widely acknowledge that the detrimental effects of climate change, in terms of ecological and economic damages, are of great concern. It is evident that policymakers are taking this evidence seriously, but the problem-solving process has proven to be more challenging than anticipated by scientists and practitioners. The 1980s saw the loss of the stratospheric ozone layer, a worldwide environmental issue with roughly similar geophysical aspects, come up for policy consideration (Schäffer, 1991). The atmospheric composition has been altered worldwide due to ozone-depleting chemical emissions, much like the climate problem. Considering the ozone case, the depletion of the stratospheric ozone layer results in heightened UV radiation, causing negative impacts on human health and agricultural production. An efficient worldwide regime was put in place in 10 years, and throughout the ensuing several decades, the ozone layer will probably return to its preindustrial state. The global ozone regime is built upon a framework convention established in 1985, followed by a protocol in 1987 and subsequent amendments (Hoel, 1991).

The remarkable achievement in global environmental policy development has generated widespread enthusiasm, leading the international community to adopt a similar approach to addressing climate change. The 1992 Framework Convention on Climate Change (UNFCCC) established a crucial objective: to stabilize GHG concentrations in the atmosphere and prevent harmful human interference with the climate system. It is important to reach this level within a reasonable time frame to enable ecosystems to naturally adapt to climate change, promote sustainable economic development, and protect food production. The 1997 Kyoto Protocol (KP) to the UNFCCC establishes specific greenhouse gas emission limits for the European Union and 37 transition economies and industrialized countries (Kinzig et al., 2013).

There has been a shift from initial optimism to prevailing pessimism, as noted by various experts. After the KP was agreed upon, it became clear that the United States, the largest GHG emitter (presently China), wouldn't be a part of the agreement. 2012, Canada withdrew from the Kyoto Protocol despite having ratified it in 2002. In addition, securing specific reduction agreements from developing countries and emerging economies has proven to be more challenging than anticipated, especially when compared to the ozone regime. Why is achieving global cooperation in addressing the climate crisis so challenging? Current studies provide empirical and systematic solutions that identify institutional structures that may facilitate problem-addressing (Elvik, 2010).

Climate change mitigation is a global communal good, frequently called an international common pool resource, whose "production" necessitates global cooperation. It is common for individuals, firms, and other actors to shift the burden of their consumption and production costs by releasing emissions into the atmosphere, which then disperse worldwide. The resulting rise in atmospheric

GHG concentrations negatively impacts everyone, although the severity varies. Addressing the issue of hazardous rates of climatic change comes with certain challenges, including considering opportunity costs for greenhouse gas emitters and the presence of a free-rider problem. Minimizing or preventing further emissions can be expensive, but the benefits extend beyond the immediate impact. The prisoner's dilemma, which also applies to the climate case as the tragedy-of-the-commons logic, poses a challenge to collective action on a global scale and global problem-solving (Morris et al., 2008).

The international system is divided into over 200 sovereign states, which implies that the issue of global collective action is framed as a political issue both within and between states. There is a significant challenge regarding global collective action between states, as the costs and benefits of problem-solving are unevenly distributed among poorer and richer countries. Greenhouse gas emissions are mainly influenced by economic output, with the ultimate objective being to separate emission trends from economic growth. It is evident that major economies are significant contributors to greenhouse gas emissions. Furthermore, aside from Brazil, India, China, and a few other developing countries, these large economies are the wealthiest and most technologically advanced nations. The imbalance in this situation has negative consequences for the possibility of worldwide collaboration (Game et al., 2014).

Countries with significant greenhouse gas emissions and substantial resources must take the lead in addressing the issue. Therefore, those who face the greatest opportunity costs are also the ones who are most likely to be least affected by climatic changes due to their strong ability to adapt. Considering the global free-rider problem, there is a clear motivation for countries to prioritize investing in adaptation measures over mitigation. This is because the country investing can directly benefit from adaptation outcomes. Developing countries, on the other hand, are more susceptible to the impacts of climatic changes due to their limited adaptation capacity. However, they would still be unable to significantly contribute to the problem's solution—apart from Brazil, India, and China—even if they stopped growing economically (Amelung and Funke, 2013).

Furthermore, compared to developed industrialized countries, those sizable emerging economies—whose dedication to significant greenhouse gas reductions is imperative—have a distinct social rate of time preference or discount rate and would rather grow first and "clean up" later. Given the unique characteristics of the climate issue within countries, it becomes even more challenging to address the global collective action problem. States often hesitate to invest in climate change mitigation, regardless of the global free-rider challenge. Discounting is a significant issue that needs to be addressed.

A significant transformation of the global energy supply system is required to achieve the necessary reduction in greenhouse gas emissions, as indicated by most climate scientists (Novriani and Surya, 2017). This transformation would involve reducing emissions to approximately 50–80% below the levels of 1990 by 2050–2100. Large-scale investments are necessary for the short and medium term, particularly in transitioning from fossil fuels to renewable energy sources. The main advantages of these investments will be seen in the long term, as they help prevent significant climate changes (Hidayat and Iksan, 2015).

Nevertheless, there is a tendency among individuals to overlook the long-term consequences of climate change and, as a result, underestimate the advantages of taking action to mitigate it. Considering the significant investments required to mitigate climate change in the short term, it leads to a relatively low net present value of such efforts. Considering the lack of significant public pressure in this regard, it is unlikely that policymakers will prioritize this issue over other matters on both international and domestic political agendas (Aini and Fathoni, 2022).

Political uncertainty poses a significant challenge to climate change mitigation efforts, as highlighted by various studies. Political instability in the climate policy realm can lead to shifting government preferences and incentives, creating uncertainty and hindering a robust long-term policy. Suppose a government is considering incentivizing private households and firms to reduce greenhouse gas emissions. These incentives could include feed-in tariffs for solar energy, subsidies, or tax breaks. If companies and individuals doubt the government's commitment to long-term climate change measures, they may hesitate to invest in mitigation efforts. This has the potential to undermine the efforts to implement the policy (Ulfah et al., 2017). This issue manifests itself both domestically and internationally. Even if a single world government had the power to autocratically decide on reducing global greenhouse gas emissions, this issue would still persist. Given the immense challenge of reducing global emissions by 80% in the next 50–70 years, it would be incredibly difficult for a hypothetical global government to make a firm and credible commitment to this course of action. It is challenging to establish a long-term credible commitment due to the frequent changes in government and the likelihood of shifting political priorities over the years (Saparwadi, 2022).

Simply put, governments and their citizens often question the value of investing billions into climate change mitigation today, fearing that a future government may "drop the ball" and allow global warming to persist. It is important to consider that altering government incentives can significantly impact ongoing costs rather than just the initial costs incurred during policy installation. It assumes that future government decisions to abandon greenhouse gas mitigation policies could reverse the progress made in climate change mitigation. This political uncertainty issue resembles the time inconsistency issue found in economic theory (Khatimah and Sugiman, 2019).

Another challenge to effectively mitigating climate change is the unequal distribution of benefits and costs within countries. Environmental problems are easier to solve when the costs of problem-solving can be spread over a very large group (small costs per actor) and the benefits of problem-solving generate large benefits for a small group of actors (large benefits per actor), as observed by Oye and Maxwell (1994), as they draw on the theories of economic regulation and collective action. This analysis sheds light on the reasons behind the relatively swift and successful resolution of the ozone problem, in contrast to the ongoing challenges faced in addressing the climate issue. Regarding the ozone case, the advantages of finding a solution were significant and could be effectively conveyed to the public. These benefits included less harm to agriculture and reduced skin cancer, as highlighted in research (Tobondo and Tondowala, 2024).

Moreover, the costs associated with resolving the issue were considerably smaller, amounting to a few billion US dollars. This starkly contrasts the climate case, where the costs are estimated to be hundreds of billions of dollars and possibly more. The issue in the ozone case was resolved by transitioning to alternative chemicals. This substitution resulted in additional economic benefits for a few major companies that dominate the global manufacturing of the relevant chemicals. Consumers were burdened with the low substitution costs per capita. As a result, the industry in question eventually accepted the suggested remedy, and the higher costs to consumers per capita were insufficient to generate sufficient opposition to halt the policy (Amalina and Vidákovich, 2023). Considering the climate situation, certain industries, such as producers of climate-friendly technologies, could gain advantages from reducing greenhouse gas emissions. However, for the average consumer and company in most economies, the costs of substituting are expected to be quite significant. This implies that significant climate policies are unlikely to be supported by the consumers or industry (several of whom also vote). However, addressing the ozone problem had a significantly higher net present value per capita than addressing the climate problem (Fitri and Abadi, 2021).

Thus, it is not shocking that there has been very little progress in negotiations for a replacement agreement for the KP since its commitment period expired in 2012. It is likely that the emission reductions of the European Community and the 37 industrialized countries, which were expected to average 5.2% (by 2012 compared to 1990), will continue to fall short of the target. On the other hand, it is worth noting that while emissions in developed nations are beginning to stabilize, they are experiencing a significant surge in developing nations (Figure 2.2). Several industrialized countries have committed to reducing greenhouse gas emissions and await an official follow-up arrangement with the KP. However, numerous studies indicate that despite the complete implementation of these pledges that lack legal obligations, there remains a significant disparity across the "emissions budget" and projected global emissions required to limit the increase in global average temperature to within 2◦ Celsius (Kusumadewi and Retnawati, 2020).

According to recent research on international regime effectiveness, it is important to consider not only emission data but also the potential outcome if there is no international cooperation, specifically in the context of the Kyoto Protocol. Research considering this factor indicates that the impact of the Kyoto Protocol on global emissions is minimal. If every unilateral reduction pledge were carried out, contingent on a follow-up to the KP, this would amount to little more than a sanguine case of business as usual. Research indicates that the 1987 Montreal Protocol, which governs certain substances that deplete the ozone layer and contribute to global warming, is proving to be more effective in reducing global warming than the KP (Nurmeidina et al., 2021).

2.2. DESIGN OF INTERNATIONAL REGIME

It is evident from the previous section that numerous challenges between and within countries make finding a global solution to environmental issues incredibly challenging. These obstacles include low net present value of GHG reductions, political uncertainty, discounting, unfavorable cost-benefit distribution, free-riding and enforcement problems, and asymmetric interests between poorer and richer countries. It also implies that studies in political science and associated fields (like political economy) have come to provide generally acknowledged justifications for the glacial pace of advancement in international climate policy. As mentioned in this section, it is important to understand the impact of different institutional design features on the global cooperative effort (Thompson, 2010).

According to (Rosendorff and Milner, 2001) the design of a regime frequently has a greater impact on effectiveness than a measure of how benign (i.e., simple to address) or malevolent (i.e., difficult to solve) the problem is. If this is the case, studies that look at the causes and effects of features of institutional architecture can be both intellectually stimulating and helpful in formulating policy. Several earlier arguments regarding addressing climate change challenges also indicate possible solutions.

2.2.1. Commitment Mechanisms and Political Uncertainty

Dealing with the issue of political uncertainty mentioned earlier necessitates the establishment of mechanisms that allow individuals, firms, and governments to make long-term commitments that are difficult and expensive to reverse in the coming decades. According to Broz (2002), there is a credibility challenge for any world government that declares a shift to a low-greenhouse gas economy in the 21st century. This is because such an investment would not yield profits for several generations, and various other issues like financial crises and poverty reduction may arise over time, making it unlikely for the ambitious climate goal to be achieved in the long run.

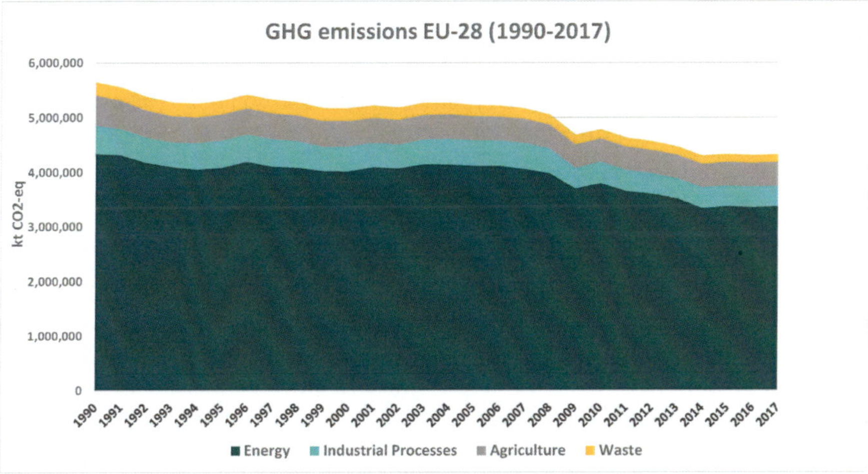

Figure 2.2. *Evolution of greenhouse gas (GHG) emissions by inventory sector in EU28.*

Source: *González-Sánchez, M.; Martín-Ortega, J.L. (2020). Greenhouse gas emissions growth in Europe: A comparative analysis of determinants. Sustainability, 12, 1012. https://doi.org/10.3390/su12031012).*

Recently, political scientists have been delving deeper into this topic. According to Morrow (1999), there is a suggestion that climate-friendly governments in major emitter countries may be motivated to establish a legally enforceable global climate agreement to limit the actions of their less climate-focused successors. This also applies to national and local government policy-makers compared to their successors. It is still uncertain and, therefore, requires further analysis to determine when governments prioritizing climate-friendly policies will have the necessary determination and ability to address the issue of political uncertainty through a binding approach. Considering the normative perspective, whether this approach is desirable is still debatable. This is because it does not consider prospective citizens' time preferences and may potentially enforce policies that are not cost-efficient or effective in solving problems (Bas and Schub, 2017).

Regardless, these arguments indicate that it is necessary to go beyond a purely economic analysis when considering the efficiency of early GHG mitigation. Considering the normative and positive analysis of political strategies and uncertainty for long-term commitments is also important. For example, normative analysis would have to make clear what factors a modern tying-hands strategy should protect against if policies change in the future and what the trade-offs between policy flexibility and rigidity are in this regard. While it may seem like a straightforward solution, addressing the issue of political uncertainty is far more complex, especially when considering the global context. Many governments often employ expensive signals to showcase their dedication to a particular policy, such as climate change. For example, the European Union clearly committed to reducing its greenhouse gas emissions by 20% by the end of 2020 (Liu et al., 2018).

Additionally, there is a clear dedication to reducing emissions even after reaching that milestone. These commitments have been incorporated into official EU regulations and programs, such as the EU's climate and energy package. However, despite the EU's strong position in the international system, several factors could potentially lead EU policymakers to reconsider their current stance and scale back or withdraw from their existing commitments (Van Ackooij et al., 2018).

If decision-making authority regarding emission quotas or levels were delegated to an independent global body or if governments legally bound themselves to a particular greenhouse gas reduction trajectory for the next few decades, a similar problem could arise. These approaches would increase the financial burden on governments if they were to backtrack in the future. However, it is possible for the government to potentially revoke the delegated authority or fail to uphold the reduction commitments made by the supranational authority (Schneider, 2020).

One potential option for institutional design is to encourage the adoption of new technologies that require significant initial investment but become more affordable over time due to network effects and scale economies. For example, as more and more vehicles are powered by electricity, hydrogen, or advanced biofuels, these vehicles and fuels may become more affordable and practical than vehicles that run on fossil fuels. Since ozone-depleting compounds were finally very simple and inexpensive to replace, global efforts to conserve the stratospheric ozone layer proved effective (Arvan, 1991). Addressing the challenges in climate policy is significantly larger in scale, necessitating reductions in emissions across a wide array of economic activities, including forestry and agricultural practices, electricity generation, and transportation. Given the scope and difficulty of the project, it is difficult to see how a technology-driven tipping point in the world's energy supply might be reached anytime soon, especially in light of the discounting issue related to high upfront costs and long-term benefits (Fink and Harms, 2012).

Without a technological solution to address the issue of political uncertainty, and considering the lack of a global governing body with approximately 200 countries involved in global climate policy-making, it is crucial to establish institutional mechanisms that can address both the issue of political uncertainty and global free-rider problems. Let's now shift the focus to this matter (Jacobs and Matthews, 2017).

2.2.2. Free Riding and Enforcement

Extensive research conducted through laboratory experiments and observational studies has consistently demonstrated the crucial role of enforcement and monitoring mechanisms in ensuring the successful provision of collective goods. This holds across various local and global scales (Konisky and Woods, 2012).

Unlike other international cooperation areas, such as trade liberalization and arms control, specific reciprocity is not an effective approach for countries seeking to address the issue of climate change. Responding to non-cooperation by one country with an increase in greenhouse gas emissions in a tit-for-tat strategy can harm both parties. This approach can potentially undermine the collective effort of providing public goods. In addition, alternative methods of decentralized enforcement, such as reducing foreign aid or implementing trade sanctions as a response to a breach of greenhouse gas reduction agreements (often referred to as diffuse or non-issue-specific reciprocity), present a secondary challenge in terms of collective goods. When a violator is punished, it may come at a cost to the punisher. However, there is a positive outcome if the punishment or the fear of it encourages the non-cooperator (and possibly others who are discouraged from violating the commitment in the future) to conform and decrease GHG emissions (Fuster and Meier, 2010).

Many believe that addressing this issue necessitates a robust centralized enforcement system. However, powerful enforcement systems are uncertain about being politically possible just when they are most required. On the other hand, when a strong enforcement system is possible from a political standpoint, it is probably unnecessary, as (Klein et al., 1988) noted. However, it could be argued that the lack of substantial advancement in global climate policymaking is not solely a result of enforcement issues but rather a consequence of the strong incentives for free-riding. Given the lack of enforcement options, countries could confidently pursue greater commitments without concerns about facing consequences for non-compliance and taking advantage of others' efforts. Several governments, including the United States, India, and China, hesitate to commit to international agreements to reduce greenhouse gas emissions. This indicates that they perceive non-compliance as having significant costs. There is some support for this assumption, as evidenced by the political backlash toward Canada's departure from the KP. Nevertheless, the significance of the enforcement and free-riding obstacle is still uncertain compared to other obstacles, including political uncertainty and discounting. Additional research could involve analyzing government justifications for non-cooperation or cooperation and conducting studies on collective goods provision (Mathew, 2017).

Regardless of the effectiveness of decentralized enforcement in global climate policy, the enforcement mechanisms of the global climate change regime are quite feeble (UNEP 2010). Of the 165 nations that have accepted the KP, only 37 have ratified specific GHG emission caps. Those 37 countries must lower emissions to 5.2% below the levels in 1990. In the event of non-compliance, certain countries are subject to sanctions that require them to reduce emissions by the required amount, along with a further 30%, in the following commitment period if they fail to meet their targets by the end of the first commitment period. In addition, countries that fail to meet their obligations are prohibited from selling emission permits in the following period. The provisions mentioned are not part of the Kyoto Protocol but are found in the Marrakesh Accord. The initial commitment period of the Kyoto Protocol concluded in late 2012, and negotiations for a subsequent period are still ongoing. Without a follow-up period, weak sanctioning mechanisms will lose relevance (Harris, 2018).

Due to the voluntary nature of engaging with the global climate regime, the issue of introducing greater enforcement to the regime presents a dilemma known as the "screening versus constraining" problem. Adding stronger enforcement measures will effectively filter out countries unable or unwilling to reduce their greenhouse gas emissions (Delmas and Keller, 2005). This will ensure that only nations committed to compliance are included. The global climate regime's design incorporates what are known as flexibility mechanisms, which reflect this trade-off. Countries with GHG reduction obligations can invest in reduction measures in foreign nations or purchase emission permits from other countries. One of the reasons behind this approach is the economic rationale, which suggests that it can help to lower mitigation costs. This is because countries with elevated marginal abatement costs can provide financial support for reductions in countries with lower marginal abatement costs. From a geophysical perspective, the location of emissions reductions is inconsequential. According to Eisenhauer (2006), those flexibility mechanisms have facilitated involvement in the global climate regime.

Several studies have explored potential solutions to the free-rider problem. Experts who have studied the options for improving the enforcement system of the KP consistently come to the same conclusion: it will be extremely difficult to do so without compromising reduction commitments or decreasing participation levels. Researchers propose alternative approaches to absolute GHG reduction targets, such as best available technology approaches, greenhouse gas intensity targets (GHG/GDP), or a more flexible system of "pledge and review" (Cirone and Urpelainen, 2013). Some critics claim that these methods would only significantly reduce overall greenhouse gas reductions. It is important to note that concluding this point would be premature. There is still much we do not know, and further research is necessary to understand better the trade-offs between different levels of participation, enforcement, and cooperation. It is still uncertain whether pursuing an inclusion strategy with softer enforcement and obligations, rather than a strategy focusing on strict enforcement and hard targets, would be more beneficial for climate policy in the long run. Additional research on trade-offs in further facets of environmental policy, along with studies on collective goods provision, might contribute to filling this research gap to some extent (Andreoni and Gee, 2012).

Efforts to shift international climate governance towards climate change adaptation and technology development can be seen as less-than-ideal solutions due to the challenges posed by discounting and global free-riding. Thanks to technological advancements, innovators can now enjoy certain benefits, even though some individuals may take advantage of the positive environmental effects without contributing themselves. Investing in climate change adaptation may bring significant returns, with the potential for the investing country to benefit greatly. Yang et al. (2018) present a thought-provoking analysis of how these incentives have shifted focus from mitigation to adaptation within the UNFCCC framework (Runge, 1984).

2.2.3. Technology and Financial Transfers and Asymmetric Interests

Given the complex nature of the international system, it is becoming increasingly challenging to mitigate greenhouse gas emissions effectively. Political uncertainty, free-rider issues, and enforcement problems among sovereign countries further complicate the situation. Additionally, the differing interests among participating countries create an additional hurdle to overcome.

Countries bear varying levels of responsibility for the current and future impacts of global warming. The distribution of such responsibilities is evolving (Irvin, 1981). Historically, the United States bears a significant responsibility for the global warming problem, which is larger than China's. However, China has emerged as the largest global emitter in recent years. In addition, there is a significant disparity in per capita emissions. While China's per capita emissions have increased somewhat, they remain a fraction of the per capita emissions in the United States (5.4 tons CO_2e/cap in China in 2008 compared to 8.0 in the EU and 18.6 in the US). Countries vary significantly in their vulnerability to and sensitivity to climatic changes (Hoekman et al., 2005).

Political science and related research disciplines have provided valuable insights into the different criteria that can be used to justify specific distributions of the global mitigation burden. Several important factors to consider are vulnerability, preferences, social rates of time, modern economic capacity, and historical responsibility. Income per capita is the only dimension on which these parameters agree rather well (Simms and Frishammar, 2024).

Like in other instances of global environmental cooperation, it will be necessary to provide financial transfers (or "green aid") from wealthier nations to less affluent ones to address imbalances of interest. Just like the stratospheric ozone situation, it is crucial to direct transfer payments toward countries with rapidly growing emissions, particularly China, India, and Brazil. However, regarding climate change, the magnitude of the challenge is significantly greater than that of the ozone case. For example, a recent estimate calculates the potential transfer payments for climate change mitigation from OECD countries to other countries. This estimate considers liability for climate change damages, differences in social rates of time preferences across regions, and a global carbon price of $35 per ton of carbon dioxide. The estimated range for these transfer payments is $15–48 billion per year (Amesse and Cohendet, 2001). Policymakers have expressed varying demands for transfer payments, ranging from $50 to $200 billion annually. It is important to note that these demands also encompass financial transfers over climate change adaptation. At least a portion of such transfers will probably be managed by the recently established Green Climate Fund based in South Korea (Eisazadeh and Abdi, 2022).

Political science literature provides valuable insights into the design and consequences of financial transfer mechanisms in various environmental policy domains. A potential limitation of this literature is its heavy reliance on a Coasian system. Transfer payments within this framework serve as a means to incentivize polluters to align their actions with the expectations of those who provide compensation to avoid pollution. If developing countries are the ones most affected by climate change, the Coasian solution may not be effective. The countries that experience the greatest harm from climate change and have the strongest motivation to address the issue are often the least financially capable. In contrast, the emerging economies and industrialized countries primarily responsible for the problem may not be willing or able to provide the necessary support. To enhance the Coasian framework, additional research should be conducted to explore non-economic motivations and incentives for financial transfers (Reichlin and Siconolfi, 1996).

Research on climate change-related financial transfers is severely limited by the lack of reliable data, which poses significant constraints. A significant issue is the duplication of development aid as climate aid, highlighted by (Roberts, 2015). Despite the challenges with data, recent research is starting to explore significant inquiries that stem from the extensive body of literature on development aid. Consider the allocation of climate aid, such as whether it is directed towards areas of greatest need or productivity or if there is a political bias (Cai et al., 2019).

Technology transfers, in addition to financial transfers, are expected to have a significant impact on the global climate regime. Political scientists have made valuable contributions to the existing body of knowledge regarding the origins and effects of specific regime designs in this area. For example, in a study by (Fiorillo et al., 2021), a game-theoretic model demonstrates the importance of rewarding technology development and adoption for a global technology fund to achieve its goals effectively. Additionally, they discovered that a fair burden-sharing regulation and a genuine commitment to funding standards are crucial (Afonso and Pinho, 2022).

2.2.4. Flexibility Mechanisms and Technical Uncertainty

Another hurdle in global climate policy arises from the lack of clarity surrounding the benefits and costs of various greenhouse gas reduction methods and the efficacy of future technologies and policy tools in addressing this issue. Technical and political uncertainty may be connected, as governments may be more inclined to break promises if the proposed policies become more expensive or less feasible than expected. However, these two types of uncertainty are distinct from each other. Efficiency and cost-effectiveness are key considerations in the global and EU climate change regime's flexibility mechanisms (Esmat et al., 2018). These mechanisms seek to steer mitigation efforts to the most effective areas and efficiently manage the uncertainty surrounding future technologies for reducing greenhouse gas emissions. The purpose of flexibility mechanisms is to reduce the costs of reducing greenhouse gas emissions. Flexibility mechanisms are valuable as they offer a more adaptable approach than regulations focusing solely on technology. This flexibility reduces the risk of countries being stuck with technology choices that may be inefficient for long-term greenhouse gas reductions. If governments can effectively carry out these functions, they may be more inclined to adopt greater reduction targets. Two flexibility mechanisms are used in the global climate regime, and they are conceptually similar (Olausson and Berggren, 2010).

A mechanism called joint implementation functions between countries that agree to greenhouse gas limits under the KP. Another mechanism, called the Clean Development Mechanism (CDM), operates between Protocol members and countries without greenhouse gas limits. Excess emission rights can be bought and sold between countries and firms, allowing those who emit less than permitted to sell their surplus to entities that exceed their allocation. By investing in emission-reduction initiatives (such as reforestation or the renovation of a coal-fired power plant) abroad, countries (or businesses) can earn credits that can be used to purchase extra emission rights domestically. Trading emission credits or allowances is a common practice among countries and firms. The European Emission Trading System (ETS) is widely recognized as this field's most crucial trading platform. The prices on this platform reflect the scarcity of greenhouse gas emission rights and the "bite" of existing climate policies (Cruz et al., 2018).

Despite the widely accepted reasons for incorporating flexibility processes in the global climate change regime, empirical research has struggled to find concrete evidence of the expected economic benefits in actual climate policy. Certain authors have utilized the rational design of institutions theory to explain the inclusion of flexibility mechanisms in the global climate regime. It has been discovered that the negotiation procedure leading to equilibrium results does not align neatly with the rational decision-making process that decision-makers are expected to follow in accomplishing these equilibria. In addition, a study by (Fink et al., 2018) indicates that governments may perceive flexibility mechanisms as a means to ease commitments and provide a safety net regarding political uncertainty, not as a technical tool to enhance cost-efficiency in reducing greenhouse gas emissions.

Much research is dedicated to examining specific flexibility mechanisms and evaluating their performance and design characteristics. Many studies in this field focus on the European ETS and CDM. The CDM plays a crucial role in the global climate regime by allowing industrialized countries to finance greenhouse gas mitigation initiatives in developing countries and count them towards their reduction commitments. For example, (Verdu et al., 2012) analyze the degree of politicization in the decision-making process of the CDM executive board. There is support for the politicization hypothesis, as it is observed that emerging nations represented on the CDM executive board have a higher likelihood of getting projects approved for their country. For example, (Srinivasan et al., 2023) contrast the ETS's enforcement mechanism with the KP's emission trading scheme and the US government's SO_2 emission trading system. They speculate whether the relative potency of enforcement tools and compliance levels are correlated. Their counterintuitive theory, which needs further empirical investigation, is that participation in the Kyoto process is voluntary. Still, the European Union's ETS is based on mandated participation, and the KP's carbon trading plan may achieve greater compliance rates (Yang et al., 2022).

Although the contracting states commonly rely on flexibility mechanisms, there is a hypothetical option that could be considered to address uncertainty: geoengineering. Various techniques in the climate context can be employed to reflect solar energy away from the atmosphere and eliminate greenhouse gases. Geoengineering could provide a flexible and efficient solution for addressing the uncertainties surrounding greenhouse gas reduction targets and the timing needed to mitigate global temperature changes. Geoengineering proposals continue to be a subject of intense debate and are unlikely to be implemented significantly shortly (Tsaousoglou et al., 2021).

Several studies conducted by political scientists have made valuable contributions to the ongoing discussion surrounding the regulation of geoengineering. There are concerns that current international treaties may not be enough to regulate geoengineering effectively. Countries can have disagreements and allow unilateral national geoengineering activities to take place. He strongly supports the implementation of geoengineering assessments and open research programs to foster an inclusive and collaborative international norm-building process. Overall, the complex nature of institutions and global policy-making results from diverse interests and the relatively recent emergence of climate change as a global issue. A contributing factor is the "band wagoning" of established international organizations that address the climate issue. According to some scholars, the emergence of this "regime complex" can be attributed to various factors such as functionality, strategy, and organization. They also argue that different political coalitions and bargaining dynamics have formed around different functional solutions and tasks. They argue that due to the political complexity and inherent uncertainty surrounding the issue of climate change, a regime complex is more inclined to be operational than a single, monolithic regime. These arguments suggest that, in general, having multiple international regimes addressing the same issue can be beneficial for problem-solving. As mentioned in a study by (Fayezi et al., 2014) regime complexes provide a potential solution in situations where one integrated governance system is not feasible. The interaction between different institutions can have both positive and synergistic outcomes and potential interference among them. Some researchers are less optimistic about the synergy between the various components of the "climate regime complex" and advocate for changes and increased centralization of systems of governance (Lecocq et al., 1998).

Lastly, some new work attempts to incorporate theories and notions from political science about the success of international regimes into holistic assessments of climate change policy. Therefore, it aims to enhance existing models that only include economic and climate components. This approach presents a potential avenue for conducting a thorough evaluation of the consequences of specific institutional design choices, as mentioned earlier, using computational simulations (Yu et al., 2023).

2.3. SUBNATIONAL AND NATIONAL CLIMATE POLICIES

Research identifying issues with international collaboration and potential solutions for regime design focuses mostly on the systemic level, i.e., all countries or groups of countries. This viewpoint is highly beneficial. However, it fails to acknowledge the significant differences among political entities in their contributions to the global collective effort of mitigating climate change. Recent political science studies have helped characterize and explain the differences in climate policies between nations and, to some extent, between subnational entities. From a normative perspective, this research suggests that the negative portrayal of the lack of global progress may be somewhat misleading. Despite the global stalemate, certain states and subnational units have determined to implement ambitious climate policies (Klemun et al., 2020).

International regimes are the result of intricate interactions among and between states. Similar to other empirical political science fields, climate politics research initially used structural characteristics to explain the behavior of subnational or national climate policies. Policy diffusion and interdependent behavior have recently been included in empirical models (Hsu et al., 2017).

Numerous studies focus on describing and evaluating climate policies at the national or regional level, particularly within the European Union. These studies provide valuable insights to the academic community and policymakers regarding the effectiveness of different policy approaches for addressing climate change. They highlight the policy instruments that have shown promising results in terms of adaptation and mitigation strategies.

It is uncommon to come across large-N comparisons of multiple countries, which are crucial for producing findings that can be applied more broadly. Many of them primarily discuss the trends or levels of greenhouse gas emissions, focusing on carbon dioxide. Instead of delving into policies, they primarily emphasize the explanation of environmental outcomes. One of the main reasons is the easy accessibility of emissions data for numerous years and countries. Many advanced studies utilize a variation of the environmental Kuznets curve (EKC) as the foundational model and subsequently incorporate political factors (Hultman et al., 2020).

The EKC has three components: income, composition, and scale. When a poor society begins to prosper economically, industrial production rises, and the average citizen puts meeting material demands ahead of protecting the environment. Environmental degradation tends to rise as incomes and economic output grow. Eventually, the curve of pollution levels off or, ideally, starts to decline. The timing of this occurrence is contingent upon a range of factors (Jörgensen et al., 2015). This happens when a higher income level leads to a greater preference for post-material goods and greater institutional, technological, and financial capacity to tackle environmental issues. Additionally, it occurs when the services sector grows while traditional pollution-intensive industries decline. After that, scale effects are outweighed by income and composition effects, and pollution decreases as income increases (Scott et al., 2023).

Several studies indicate that the turning point on the EKC for carbon dioxide occurs at relatively high-income levels. The rationale is that political priorities are given to pollutant types that cause immediate, localized harm instead of those that can be "exported" into the atmosphere and have indirect long-term effects. It is worth noting that the turning points of certain types of water and air pollution often happen at lower income levels than greenhouse gas emissions. Research also indicates that countries' greenhouse gas emissions vary significantly, even considering their income levels (Fisher, 2013).

Political scientists have been studying the factors contributing to emission variations, particularly focusing on good governance indicators and democracy. Many studies fail to find these factors' strong and significant impact. This finding contradicts the findings of other studies that have observed a relationship between democracy and local forms of pollution, specifically air pollution. GHG emissions are primarily influenced by socioeconomic factors such as the historical development of transportation networks, past choices in energy supply, industrial structure, population density, and income levels. The impact of democracy on GHG emissions seems to be negligible. It aligns with the belief that when it comes to large-scale and long-term environmental issues, free-rider and discounting problems tend to outweigh the impact of democracy. Unfortunately, this result may present a misleadingly positive view of advanced industrialized democracies due to outsourcing pollution-intensive economic operations to less developed nations. It's interesting to note, nevertheless, that

(Happaerts, 2015) discovers that, regardless of their degree of democracy, developing nations with more membership in international organizations typically have lower greenhouse gas emissions. The mechanisms by which this involvement affects greenhouse gas emissions still need to be clarified. However, Spilker's findings imply that this aspect can aid in reducing the "grow rich and clean up later" issue, regardless of national characteristics that may affect membership in international organizations (Xu, 2021).

Various studies examining climate policies, specifically their impact on democracy, have yielded diverse findings. Many individuals focus on ratifying the Kyoto Protocol, which significantly indicates their commitment to environmental policy. For example, (Jordaan et al., 2019) view the participation of European countries in the KP as a form of "Europeanization." This refers to the trend of EU countries ratifying the agreement together, with leading countries motivating the slower ones through various incentives, such as the possibility of EU accession. According to other research, democracies are more likely than nations to implement more aggressive climate policy commitments (Ceteris paribus). Another study by highlights the valuable insights that can be gained from analyzing the reconstruction of individual countries' ratification decisions. Through a series of interviews, this study delves into the discrepancy between the commonly accepted two-level-game argument and the decision of the US not to participate in the KP (Caron et al., 2015). They argue that the Clinton-Gore administration had already given up on Senate ratification when the KP was finalized. However, they still pushed for the agreement to establish a "green legacy" for themselves.

A significant amount of research still needs to be conducted in this field to draw reliable conclusions regarding the factors that contribute to differences in the levels and types of ambition of climate policies among different political units. Considering political units' interconnectedness and unique characteristics, this work must address policy diffusion. For this kind of research, having more accurate and comprehensive information regarding climate policies, especially panel data, is crucial. The latter type of data structure is essential in empirically modeling interdependent political unit decision-making. Currently, only two datasets are available at the international level, one by German watch and the other by (Rodríguez et al., 2022). The datasets provide overall measures of the ambition levels for climate policies but do not include details on the specific climate policy instruments that have been implemented. Current work is being done on subnational climate policies to provide panel datasets.

Based on my expertise in environmental policy, it is evident from recent research that there is a significant level of interdependent behavior among political units regarding national and local climate policy. For example, a study analyzed the factors that influence the involvement of American cities in the Mayor's Climate Protection Agreement (Bieler et al., 2017). According to her research, the main factors influencing these decisions are the economic structure of the city, political inclination, educational attainment, population, and whether or not nearby communities ratify the agreement. When examining various interpretations of the neighborhood, (Fuhr et al., 2018) discover that interdependent decision-making does not appear to happen through direct geographical neighborhood connections. Instead, it occurs in cities with similar population sizes and their involvement in municipal networks. According to a recent study, federal climate programs in the US tend to focus on intangible reputational benefits, while state programs prioritize more concrete benefits like access to capital and technical knowledge (de Macedo and Jacobi, 2019). This finding raises questions about the impact of horizontal policy diffusion on vertical policy diffusion, which still needs to be addressed (Valenzuela, 2014).

Further research focuses on the global perspective. They analyze the spread of specific climate policy technologies or instruments across different countries using qualitative case studies. Furthermore, additional research has shed light on conceptualizing how climate policy spreads within federal systems and across international borders (Anderton and Setzer, 2018).

It remains to be seen through future research how climate policy diffusion mechanisms between states compare to those within states. Additionally, it is important to consider the complex interplay between international and domestic factors when analyzing policy choices. One instance of such research focusing on emissions rather than policies is a study (Lamb and Minx, 2020). It has been observed that, in specific circumstances, foreign direct investment has the potential to decrease the carbon intensity of economic activity in the countries receiving it. Another study conducted explores the impact of a conditional commitment by a significant polluter on the willingness of different nations to cooperate, taking into account the implications for mitigation costs dependent on domestic politics (Jänicke and Quitzow, 2017). According to their findings, it seems unlikely that Europe's progress in climate policy will inspire the US to do the same. A limited number of studies analyze the spread of international climate policies, specifically focusing on taxation. One such study is conducted by Dubash et al., (2013). Examining temporal and spatial variation in "green taxes" as a function of international and domestic determinants, the focus is on OECD countries from 1995 — 2004. It has been observed that within the country, the positions of the legislative medians on the left-right spectrum and the influence of the energy-producing sector have the greatest impact. National green taxes are also influenced internationally by commerce and global environmental governance networks.

Research considering international and domestic factors influencing climate policy can shed light on why certain subnational units or countries emerge as leaders despite the challenges posed by global free-rider problems and other systemic barriers. They will also assist in addressing the concerns of those who claim that taking the lead in climate change is only a temporary trend that relies on easily achievable benefits, such as creating jobs in the renewable energy industry or reducing local air pollution.

Based on my expertise in environmental policy, it is evident from the literature that greenhouse gas emissions are predominantly influenced by economic processes and structures rather than political factors like good governance and democracy. Climate policy ambition levels and forms, which can indicate future emissions behavior, are influenced by political system characteristics, political network effects, and interdependent behavior (Walters, 2006).

Ultimately, the development of comprehensive and more detailed datasets on global climate policies and the utilization of advanced spatial statistics for data analysis will continue to be a work in progress. There are many nuances in climate policies that numerical coding and statistical techniques cannot capture. These techniques cannot fully capture the dynamic processes of decision-making that lead to the different policy choices we observe. Therefore, in-depth case studies that provide detailed insights into climate policy-making will be essential. In this context, it is important to acknowledge that political science is experiencing a rise in quantitative research on climate change policy. This is not to say that quantitative work is inherently better, but rather that there is a need to explore the generalizability of findings from extensive qualitative research through large-scale statistical analysis.

2.4. BEYOND THE STATE

Research on the climate change issue in political science has predominantly concentrated on government behavior and the institutional components of climate politics. Given the need for a complete overhaul of the carbon-based energy system on a global scale, it is crucial to garner public support for climate policy. This is because such a solution will significantly affect individuals and businesses. Research in the field of public opinion, the science-policy interface, and climate governance is driven by this important consideration. It is argued that the traditional method of state consent to international treaties is no longer enough to legitimize complex global policy challenges which are regulated and negotiated far from the domestic audience (Rose and Miller, 2010).

2.4.1. Civil Society

Climate politics is marked by a significant level of engagement from various sectors of society. During recent UNFCCC Conferences, many conference participants hailed from NGOs, including environmental groups, scientific institutions, and business and environmental groups. Remarkably, around 70% of national delegations had nongovernmental representatives, with approximately 18% of delegation members coming from NGOs (Chambers and Kopstein, 2001).

The literature on civil society involvement in global climate governance provides valuable interpretations, typologies, and descriptions of this phenomenon. However, it still lacks comprehensive solutions to certain fundamental inquiries. For example, governments often include civil society in various ways, although the specific practices vary greatly among governments. Additionally, it is important to understand the circumstances under which civil society involvement has specific impacts on governance initiatives (Kaldor, 2003).

2.4.2. Public Opinion

As mentioned earlier, two significant challenges in climate politics are the tendency to undervalue the long-term benefits of implementing more stringent climate policies now and the unequal distribution of benefits and costs of climate policy both within and between nations. In this context, public opinion holds a significant influence over climate politics (Burstein, 2003).

Research on public opinion regarding climate policy is still in the early stages, as demonstrated by previous studies on climate risk perceptions and environmental attitudes in psychology and sociology. However, this area has potential for further development, as evidenced by recent research. Interesting additions to current research deal with both specific aspects of climate politics, including civil society involvement and reciprocity, and broader concerns, like how the weather affects people's perceptions of climate risk and how framing affects policy preferences.

Regarding broader concerns, Gamson (2004) investigated the potential impact of above-normal temperatures on individuals' perceptions of global warming. The two authors discover that above-normal temperatures lead individuals to conclude that there is strong evidence for global warming, which is consistent with the widely held belief among climate researchers that people tend to conflate the notions of climate and weather. This impact is temporary, though it is more pronounced among individuals with lower levels of education. The analysis relies on data from the US. This result is consistent with past studies on American views on climate change. For example, (Rokeach, 1968) discovered that elite cues and economic conditions significantly impact climate policy preferences more than other factors like scientific risk communication. According to their findings, the impact of information-based science advocacy on public concern about climate change has been relatively insignificant. Instead, they argue that political mobilization by advocacy groups and elites plays a crucial role in shaping public attitudes toward climate change. In another study, investigate how the way policy decisions are framed may influence Australians who oppose climate change to support more stringent regulations (Page et al., 1987). It has been observed that individuals who are skeptical about climate change tend to show more support when climate policies are presented as promoting economic progress and fostering a stronger sense of community. On the other hand, presenting information regarding the potential consequences of climatic changes does not yield any results (O'Lear, 2016).

Studies on particular aspects of international climate politics have also been conducted. Through survey experiments, (Lax and Phillips, 2009) explore the potential for promoting global climate cooperation through reciprocity strategies. It has been observed that people do not perceive climate politics in terms of tit-for-tat exchanges. This finding aligns with the commonly held belief, discussed earlier, that climate governance relying on strict reciprocity is improbable. According to Burstein, (1998), respondents expressed their support for economic sanctions against countries that breach an international climate agreement. They comment that cooperation may arise from connecting climate with other matters and incorporating climate commitments into international law. In a study, survey experiments were utilized to examine the potential impact of increased civil society involvement on public support for climate policy (Osborne and Rose, 1999). The inclusion of representatives from civil society in the process of formulating climate policies is widely supported. It has been observed that including civil society in global climate policymaking leads to more positive assessments (Hsiang et al., 2013).

In comparison, the exclusion of civil society leads to more negative assessments. These findings indirectly answer the earlier question of why governments engage civil society in global climate policy-making. In conclusion, it is important to highlight that a significant portion of the research on public opinion regarding climate change focuses on understanding risk and problem perceptions. This includes studies in political science journals and authored by political scientists. Studies have not been able to explain individual attitudes or behavior concerning ambition levels and forms of climate policy, with very few exceptions (Chong and Druckman, 2010).

2.5. HUMAN SECURITY, POLITICAL VIOLENCE, AND CLIMATE CHANGE

It is evident from the previous sections that achieving a stable level of greenhouse gas emissions to prevent significant alterations in the Earth's climate system will be a highly demanding task. What are the potential consequences if political efforts are unsuccessful? Extensive research has demonstrated the significant impact of climatic changes on people's livelihoods and national economies. Less developed countries will likely face the most severe consequences, especially those in semi-arid or arid zones with high-population coastal areas. According to the Stern Report, a comprehensive analysis of climate change prevention, it has been determined that if severe climate change occurs with a temperature increase of 3–6 degrees, the economic losses could amount to 5–20% of GDP annually. Other economic evaluations have reached smaller figures (Barnett and Adger, 2007).

It is widely believed that the significant impact of severe climatic changes on people's livelihoods and economic systems can potentially escalate the risk of political violence, even leading to war. For example, Kofi Annan, the former UN Secretary-General, emphasized that climate change poses a significant threat to global peace and security (De Rivera and Páez, 2007). The IPCC, a global organization of scientists, has supported these claims in its 2007 report. They regularly gather and evaluate scientific knowledge on climate change, including its causes, consequences, and possible solutions (Brown et al., 2007).

Given the pressing concerns surrounding the political consequences of unaddressed climate change, a collaborative research initiative has emerged, uniting environmental policy experts and conflict researchers. Throughout the 1980s and 1990s, there was a resurgence of an argument originally made by Thomas Malthus in 1798, following the "limits to growth" debate. Various studies show a correlation between political violence and environmental degradation. Through extensive research, many causal pathways have been identified at the conceptual level, illustrating how environmental degradation can lead to violent conflict. Several qualitative case studies have supported such claims (Iqbal, 2006).

While this literature indicates that environmental degradation can potentially lead to conflict, including violent conflict, it does have a few limitations. For example, the cases examined were not randomly selected and tended to concentrate on small-scale conflicts. These characteristics make it challenging to draw firm conclusions regarding the conditions under which environmental degradation causes specific kinds of violent conflict and why it does not. Several recent studies have successfully addressed a few of the limitations in this field through large-scale quantitative analysis. Many studies have examined the potential link between climatic changes and the risk of violent conflict. They have primarily focused on addressing the claim of climate-conflict from three different perspectives (Levy et al., 2017).

They have analyzed whether there is a direct correlation between climatic variability (or climate changes) and huge-scale political violence, such as civil or interstate war, as studied by (Mack, 2005); Secondly, they have researched possible indirect impacts to align empirical testing closer with theoretical arguments. It has been argued that the risk of violence may be influenced by climate change due to its impact on economic activity. For example, recent studies have investigated the potential link between climate change and political violence, specifically looking at its potential negative effects on economic growth (Exenberger and Pondorfer, 2013). Despite initial claims to the contrary, the results have overwhelmingly shown no robust evidence that climatic changes are consistently linked with large-scale violent conflict. The special issue of the Journal of Peace Research 49/1, 2012 is a turning point in this research. This lack of results is significant, as it questions a widely made political assertion that has often been stated and has indirectly supported the need for more robust climate policies. Thirdly, the previous literature on environmental conflicts lacked clarity in defining conflict (Vestby, 2019). However, quantitative research has compelled researchers to provide more explicit explanations for the outcomes they are studying. This applies to both the intensity and scale. Based on my expertise in environmental policy, it is important to note that the research mentioned above indicates that the likelihood of climatic changes causing significant political violence on a large scale, such as civil war or interstate war, is minimal. This finding remains consistent even when using spatially disaggregated civil war data. Unfortunately, it does not provide any insights into whether climate change has led to smaller or less intense conflicts (Scheffran and Battaglini, 2011).

It is worth noting that recent studies utilizing new event datasets have produced conflicting findings regarding lower-intensity conflict. For example, a study reveals a correlation between political conflict and African rainfall variability (Von Uexkull and Buhaug, 2021). However, it is worth noting that violent events are more likely to occur in wetter years. Both extreme deviations in rainfall, whether wetter or drier, can lead to an increase in both nonviolent and violent political conflict. Based on my expertise in environmental policy, it is worth noting that a recent study focused on water-related conflicts, which are often associated with climate change. Surprisingly, the study reveals that violent water-related conflicts, even at small local levels, are rare (Raleigh and Urdal, 2007).

On the contrary, the study highlights that cooperation is quite common. Furthermore, the research indicates that water conflicts are mainly caused by the increasing demand for water rather than any reductions in water supply due to climate change (Ide et al., 2014). Based on extensive case studies on local resources management, this research indicates that standardized conflict resolution and adaptation mechanisms are key factors in preventing violent water-related conflicts (Brzoska and Fröhlich, 2016).

The research presented here sheds light on possible drivers of policy aimed at mitigating climate change. Naturally, it cannot demonstrate the potential outcomes in the face of severe climate change scenarios. It implies, though, that the primary justifications for greenhouse gas mitigation policies should be focused on the various negative consequences of climate change that lead to human suffering. It is important to consider justifications based on economic damages and human security when making policy decisions (Koren and Schon, 2023). The idea of human security gained significant political attention following the release of the UNDP 1994 Human Development Report. It prioritizes the individual's well-being over the state and encompasses various aspects of human welfare, including human rights, health, food security, and employment. While it seems that political science research is progressing in resolving the climate war debate, there are still significant research opportunities to explore the impacts of climate change on instability and low-intensity political violence, adaptation strategies, migration, and other aspects of human security (Livingstone, 2015).

2.6. SUMMARY

The substantive questions and techniques of political science research on climate change issues vary remarkably. Its subjects cover anything from public opinion to establishing global regimes to the sociopolitical effects of climate change. It utilizes various modern social science methods, including experimental approaches, statistical work, case studies, and formal and verbal theory.

The wide range of research areas in political science, such as electoral behavior, democratization, and war, contrasts with the focus on climate change. Unlike these other areas, political science studies on climate change do not limit themselves to specific outcome variables. Instead, climate change is seen as a complex issue that political scientists and scholars from various disciplines are working to understand using their analytical tools. Many political science studies on climate change are problem-oriented, which makes their findings important to policy. However, an integrated research community has not yet emerged. Therefore, a significant portion of the most groundbreaking research on climate politics has been published in journals outside the field of political science. While policy-makers might not prioritize this issue, the field of political science must do so.

As mentioned earlier, numerous thought-provoking and captivating research questions must be tackled. Considering the sluggish progress of global governance efforts, focusing more on the dynamics at lower levels would be beneficial. This would provide valuable insights for analysis, as there is a significant amount of variation to be understood. Additionally, from a normative standpoint, giving greater attention to these lower-level dynamics is important.

The combination of a significant emphasis on current mitigation costs and an intense discounting of future benefits of averting climate change-related harms represent one of the most fundamental barriers to climate policy at the micro level. Further investigation is necessary to explore how different approaches to presenting potential solutions could minimize discounting and enhance public backing for policies aimed at mitigating climate change. Previous environmental issues, like water quality, ozone layer depletion, and leaded gasoline, have shown that effective policies can yield positive results when policymakers effectively communicate the significant long-term benefits to the public and garner support from influential stakeholders. Research in social psychology indicates that effective framing may depend on more than just rationalistic arguments, such as reducing risks and the benefits of mitigation policies. It also suggests that identity frames and sense-of-community play a crucial role.

When considering the meso level, there are significant concerns regarding civil society participation and the connection between policy and science. There is a need for further research on the impact of increased civil society involvement in global to local climate governance on these efforts' process and outcome qualities. Engaging civil society is widely believed to be advantageous as it brings in specialized knowledge and enhances public support. Nevertheless, further investigation is needed to determine the specific circumstances in which this is true. It is interesting to note that the scientific evidence on climate risks and mitigation possibilities is interpreted and discussed in various ways within different political systems. It is crucial to understand the factors shaping how political leaders interpret scientific data and make decisions based on that information. Another crucial question remains unanswered: how can climate-friendly governments at various levels effectively respond to political uncertainty by implementing policies that future administrations cannot easily reverse? A particularly attractive place to start for this kind of research is ongoing research around environmental innovation and target group influence (Janicke 2012, Skodvin et al., 2010). It highlights the focus on disseminating climate-friendly policy instruments and policies among subnational units, states, and cities and how these local processes relate to international governance efforts in this field. Given the challenges of mobilizing sufficient transfer payments to address emerging economies, it becomes crucial to prioritize research at the meso and micro levels in such nations.

From a broader perspective, there is still an unanswered question regarding the extent of the challenges posed by free-riding and enforcement issues in global climate governance. Considering the trade-offs between different approaches to environmental policy, it is essential to examine the balance between technology oriented solutions and stricter obligations centered on enforcement mechanisms and emission targets. This analysis becomes particularly important when assessing their impact on participation. In addition, there is still limited understanding of the effectiveness of the flexibility mechanisms in regional and global climate governance systems in reducing mitigation costs and encouraging states to make more ambitious commitments. It remains uncertain whether these mechanisms are designed to facilitate engagement genuinely or if they are primarily used to provide a way to increase participation and ease strict obligations. Another crucial issue that has received limited attention in political science research is the impact of unilateral policies on international climate governance. In other environmental areas, there have been instances where certain jurisdictions have taken the lead and influenced others to follow through with a mix of incentives and consequences. The EU has recently implemented a "cap and trade" system for all airlines operating flights in, from, and to Europe in climate policy. Several governments outside the EU, including the US, India, and China, have expressed their objections to this measure but have not taken retaliatory actions. Examining if this new policy can inspire non-EU airlines to decrease their emissions will be intriguing. Similarly, border carbon adjustment tariffs have been suggested, but they have not been put into action. Lastly, it is crucial to research climate change's impacts on adaptation strategies, migration patterns, instability and low-intensity political violence, and various aspects of human security.

REFERENCES

1. Afonso, O., & Pinho, M. (2022). How to reverse a negative asymmetric labor productivity shock in the European Union? A directed technical change analysis with fiscal and monetary policies. *Mathematical Social Sciences, 116*, 47–67.
2. Aini, H. N., & Fathoni, A. (2022). Pengembangan Lembar Kerja Peserta Didik (LKPD) Matematika Berbasis Budaya Lokal Siswa Sekolah Dasar. *Jurnal Basicedu, 6*(4), 6167–6174.
3. Amalina, I. K., & Vidákovich, T. (2023). Assessment of domain-specific prior knowledge: A development and validation of mathematical problem-solving test. *International Journal of Evaluation and Research in Education, 12*(1), 468–476.
4. Amelung, D., & Funke, J. (2013). Dealing with the uncertainties of climate engineering: Warnings from a psychological complex problem-solving perspective. *Technology in Society, 35*(1), 32–40.
5. Amesse, F., & Cohendet, P. (2001). Technology transfer revisited from the perspective of the knowledge-based economy. *Research Policy, 30*(9), 1459–1478.
6. Anderton, K., & Setzer, J. (2018). Subnational climate entrepreneurship: Innovative climate action in California and São Paulo. *Regional Environmental Change, 18*, 1273–1284.
7. Andreoni, J., & Gee, L. K. (2012). Gun for hire: Delegated enforcement and peer punishment in public goods provision. *Journal of Public Economics, 96*(11, 12), 1036–1046.
8. Arvan, L. (1991). Flexibility versus commitment in strategic trade policy under uncertainty: A model of endogenous policy leadership. *Journal of International Economics, 31*(3, 4), 341–355.
9. Barnett, J., & Adger, W. N. (2007). Climate change, human security, and violent conflict. *Political Geography, 26*(6), 639–655.
10. Bas, M. A., & Schub, R. (2017). Peaceful uncertainty: When power shocks do not create commitment problems. *International Studies Quarterly, 61*(4), 850–866.
11. Bieler, A., Haluza-Delay, R., Dale, A., & McKenzie, M. (2017). A national overview of climate change education policy: Policy coherence between subnational climate and education policies in Canada (K-12). *Journal of Education for Sustainable Development, 11*(2), 63–85.
12. Brown, O., Hammill, A., & McLeman, R. (2007). Climate change as the

'new' security threat: Implications for Africa. *International Affairs, 83*(6), 1141–1154.

13. Broz, J. L. (2002). Political system transparency and monetary commitment regimes. *International Organization, 56*(4), 861–887.
14. Brzoska, M., & Fröhlich, C. (2016). Climate change, migration, and violent conflict: Vulnerabilities, pathways, and adaptation strategies. *Migration and Development, 5*(2), 190–210.
15. Burstein, P. (1998). Bringing the public back in: Should sociologists consider the impact of public opinion on public policy? *Social Forces, 77*(1), 27–62.
16. Burstein, P. (2003). The impact of public opinion on public policy: A review and an agenda. *Political Research Quarterly, 56*(1), 29–40.
17. Cai, D., Guo, C., & Tan, Y. (2019). Design of incentive contract for technological innovation of new energy vehicles with asymmetric information. *International Journal of Environmental Research and Public Health, 16*(22), 4544.
18. Caron, J., Rausch, S., & Winchester, N. (2015). Leakage from sub-national climate policy: The case of California's cap-and-trade program. *The Energy Journal, 36*(2), 167–190.
19. Chambers, S., & Kopstein, J. (2001). Bad civil society. *Political Theory, 29*(6), 837–865.
20. Chong, D., & Druckman, J. N. (2010). Dynamic public opinion: Communication effects over time. *American Political Science Review, 104*(4), 663–680.
21. Cirone, A. E., & Urpelainen, J. (2013). Trade sanctions in international environmental policy: Deterring or encouraging free riding? *Conflict Management and Peace Science, 30*(4), 309–334.
22. Cruz, M. R., Fitiwi, D. Z., Santos, S. F., & Catalão, J. P. (2018). A comprehensive survey of flexibility options for supporting the low-carbon energy future. *Renewable and Sustainable Energy Reviews, 97*, 338–353.
23. de Macedo, L. S. V., & Jacobi, P. R. (2019). Subnational politics of the urban age: Evidence from Brazil on integrating global climate goals in the municipal agenda. *Palgrave Communications, 5*(1), 1–15.
24. De Rivera, J., & Páez, D. (2007). Emotional climate, human security, and cultures of peace. *Journal of Social Issues, 63*(2), 233–253.
25. Delmas, M., & Keller, A. (2005). Free riding in voluntary environmental programs: The case of the US EPA WasteWise program. *Policy Sciences, 38*(2), 91–106.

26. Dubash, N. K., Hagemann, M., Höhne, N., & Upadhyaya, P. (2013). Developments in national climate change mitigation legislation and strategy. *Climate Policy, 13*(6), 649–664.
27. Eisazadeh Roshan, Y., & Abdi, Y. (2022). ICT and information asymmetry; New evidence of the financial system in selected MENA countries. *Iranian Economic Review, 26*(2), 445–458.
28. Eisenhauer, J. G. (2006). Conscience as a deterrent to free riding. *International Journal of Social Economics, 33*(8), 534–546.
29. Elvik, R. (2010). Why some road safety problems are more difficult to solve than others. *Accident Analysis & Prevention, 42*(4), 1089–1096.
30. Esmat, A., Usaola, J., & Moreno, M. Á. (2018). A decentralized local flexibility market considering the uncertainty of demand. *Energies, 11*(8), 2078.
31. Exenberger, A., & Pondorfer, A. (2013). Climate change and the risk of mass violence: Africa in the 21st century. *Peace Economics, Peace Science and Public Policy, 19*(3), 381–392.
32. Fayezi, S., Zutshi, A., & O'Loughlin, A. (2014). Developing an analytical framework to assess the uncertainty and flexibility mismatches across the supply chain. *Business Process Management Journal, 20*(3), 362–391.
33. Fink, L., Wyss, S., & Lichtenstein, Y. (2018). Aligning flexibility with uncertainty in software development arrangements through a contractual typology. *Journal of Global Operations and Strategic Sourcing, 11*(1), 2–26.
34. Fink, M., & Harms, R. (2012). Contextualizing the relationship between self-commitment and performance: Environmental and behavioural uncertainty in (cross-border) alliances of SMEs. *Entrepreneurship & Regional Development, 24*(3, 4), 161–179.
35. Fiorillo, F., Giuranno, M. G., & Sacchi, A. (2021). Asymmetric decentralization: Distortions and opportunities. *Economia Politica, 38*, 625–656.
36. Fisher, D. R. (2013). Understanding the relationship between subnational and national climate change politics in the United States: Toward a theory of boomerang federalism. *Environment and Planning C: Government and Policy, 31*(5), 769–784.
37. Fitri, A., & Abadi, A. M. (2021). Kesulitan siswa SMA dalam menyelesaikan soal matematika pada materi peluang. *Jurnal Riset Pendidikan Matematika, 8*(1), 96–105.
38. Fuhr, H., Hickmann, T., & Kern, K. (2018). The role of cities in multi-

level climate governance: Local climate policies and the 1.5 C target. *Current Opinion in Environmental Sustainability, 30*, 1–6.

39. Fuster, A., & Meier, S. (2010). Another hidden cost of incentives: The detrimental effect on norm enforcement. *Management Science, 56*(1), 57–70.
40. Game, E. T., Meijaard, E., Sheil, D., & McDonald Madden, E. (2014). Conservation in a wicked complex world; Challenges and solutions. *Conservation Letters, 7*(3), 271–277.
41. Gamson, W. A. (2004). Bystanders, public opinion, and the media. In D. A. Snow, S. A. Soule, & H. Kriesi (Eds.), *The Blackwell Companion to Social Movements* (pp. 242–261). Blackwell Publishing.
42. Gowdy, J. M. (2008). Behavioral economics and climate change policy. *Journal of Economic Behavior & Organization, 68*(3, 4), 632–644.
43. Grubb, M. (2004). Technology innovation and climate change policy: An overview of issues and options. *Keio Economic Studies, 41*(2), 103–132.
44. Gupta, J. (2010). A history of international climate change policy. *Wiley Interdisciplinary Reviews: Climate Change, 1*(5), 636–653.
45. Happaerts, S. (2015). Climate governance in federal Belgium: Modest subnational policies in a complex multi-level setting. *Journal of Integrative Environmental Sciences, 12*(4), 285–301.
46. Harris, C. (2018). Institutional solutions to free-riding in peer-to-peer networks: A case study of online pirate communities. *Journal of Institutional Economics, 14*(5), 901–924.
47. Hidayat, R., & Iksan, Z. H. (2015). The effect of realistic mathematic education on students' conceptual understanding of linear programming. *Creative Education, 6*(22), 2438–2445.
48. Hoekman, B. M., Maskus, K. E., & Saggi, K. (2005). Transfer of technology to developing countries: Unilateral and multilateral policy options. *World Development, 33*(10), 1587–1602.
49. Hoel, M. (1991). Global environmental problems: The effects of unilateral actions taken by one country. *Journal of Environmental Economics and Management, 20*(1), 55–70.
50. Hsiang, S. M., Burke, M., & Miguel, E. (2013). Quantifying the influence of climate on human conflict. *Science, 341*(6151), 1235367.
51. Hsu, A., Weinfurter, A. J., & Xu, K. (2017). Aligning subnational climate actions for the new post-Paris climate regime. *Climatic Change, 142*, 419–432.
52. Hultman, N. E., Clarke, L., Frisch, C., Kennedy, K., McJeon, H., Cyrs,

T., & O'Neill, J. (2020). Fusing subnational with national climate action is central to decarbonization: The case of the United States. *Nature Communications, 11*(1), 5255.

53. Ide, T., Schilling, J., Link, J. S., Scheffran, J., Ngaruiya, G., & Weinzierl, T. (2014). On exposure, vulnerability and violence: Spatial distribution of risk factors for climate change and violent conflict across Kenya and Uganda. *Political Geography, 43*, 68–81.
54. Iqbal, Z. (2006). Health and human security: The public health impact of violent conflict. *International Studies Quarterly, 50*(3), 631–649.
55. Irvin, G. W. (1981). Bargaining asymmetry in technology transfer: A game theoretical approach. *The Journal of Development Studies, 18*(1), 85–93.
56. Jacobs, A. M., & Matthews, J. S. (2017). Policy attitudes in institutional context: Rules, uncertainty, and the mass politics of public investment. *American Journal of Political Science, 61*(1), 194–207.
57. Jänicke, M., & Quitzow, R. (2017). Multi-level reinforcement in European climate and energy governance: Mobilizing economic interests at the sub-national levels. *Environmental Policy and Governance, 27*(2), 122–136.
58. Jordaan, S. M., Davidson, A., Nazari, J. A., & Herremans, I. M. (2019). The dynamics of advancing climate policy in federal political systems. *Environmental Policy and Governance, 29*(3), 220–234.
59. Jörgensen, K., Jogesh, A., & Mishra, A. (2015). Multi-level climate governance and the role of the subnational level. *Journal of Integrative Environmental Sciences, 12*(4), 235–245.
60. Kaldor, M. (2003). Civil society and accountability. *Journal of Human Development, 4*(1), 5–27.
61. Khatimah, H., & Sugiman, S. (2019). The effect of problem-solving approach to mathematics problem-solving ability in fifth grade. *Journal of Physics: Conference Series, 1157*(4), 042104.
62. Kinzig, A. P., Ehrlich, P. R., Alston, L. J., Arrow, K., Barrett, S., Buchman, T. G., & Saari, D. (2013). Social norms and global environmental challenges: The complex interaction of behaviors, values, and policy. *BioScience, 63*(3), 164–175.
63. Klein, B., & Murphy, K. M. (1988). Vertical restraints as contract enforcement mechanisms. *The Journal of Law and Economics, 31*(2), 265–297.
64. Klemun, M. M., Edwards, M. R., & Trancik, J. E. (2020). Research priorities for supporting subnational climate policies. *Wiley Interdisciplinary Reviews: Climate Change, 11*(6), e646.

65. Konisky, D. M., & Woods, N. D. (2012). Environmental free riding in state water pollution enforcement. *State Politics & Policy Quarterly, 12*(3), 227–251.
66. Koren, O., & Schon, J. (2023). Climate change, cash crops, and violence against civilians in the Sahel. *Regional Environmental Change, 23*(3), 112.
67. Kusumadewi, C. A., & Retnawati, H. (2020). Identification of elementary school students' difficulties in mathematical problem-solving. *Journal of Physics: Conference Series, 1511*(1), 012031.
68. Lacey, J., Howden, M., Cvitanovic, C., & Colvin, R. M. (2018). Understanding and managing trust at the climate science–policy interface. *Nature Climate Change, 8*(1), 22–28.
69. Lamb, W. F., & Minx, J. C. (2020). The political economy of national climate policy: Architectures of constraint and a typology of countries. *Energy Research & Social Science, 64*, 101429.
70. Lax, J. R., & Phillips, J. H. (2009). How should we estimate public opinion in the states? *American Journal of Political Science, 53*(1), 107–121.
71. Lecocq, F., Hourcade, J. C., & Duong, M. H. (1998). Decision making under uncertainty and inertia constraints: Sectoral implications of the when flexibility. *Energy Economics, 20*(5, 6), 539–555.
72. Levy, B. S., Sidel, V. W., & Patz, J. A. (2017). Climate change and collective violence. *Annual Review of Public Health, 38*(1), 241–257.
73. Liu, N., Tang, S. Y., Zhan, X., & Lo, C. W. H. (2018). Political commitment, policy ambiguity, and corporate environmental practices. *Policy Studies Journal, 46*(1), 190–214.
74. Livingstone, D. N. (2015). The climate of war: Violence, warfare, and climatic reductionism. *Wiley Interdisciplinary Reviews: Climate Change, 6*(5), 437–444.
75. Mack, A. (2005). Human security report 2005: War and peace in the 21st century. *Die Friedens-Warte, 80*(1/2), 177–191.
76. Mathew, S. (2017). How the second-order free rider problem is solved in a small-scale society. *American Economic Review, 107*(5), 578–581.
77. Morris, S. S., Cogill, B., & Uauy, R. (2008). Effective international action against undernutrition: Why has it proven so difficult and what can be done to accelerate progress? *The Lancet, 371*(9612), 608–621.
78. Morrow, J. D. (1999). The strategic setting of choices: Signaling, commitment, and negotiation in international politics. In D. A. Lake & R.

Powell (Eds.), *Strategic Choice and International Relations* (pp. 86–91). Princeton University Press.

79. Novriani, M. R., & Surya, E. (2017). Analysis of student difficulties in mathematics problem solving ability at MTs SWASTA IRA Medan. *International Journal of Sciences: Basic and Applied Research (IJSBAR), 33*(3), 63–75.
80. Nurmeidina, R., Lazwardi, A., & Nugroho, A. G. (2021). Pengembangan modul trigonometri untuk mengembangkan kemampuan pemecahan masalah matematika. *Aksioma: Jurnal Program Studi Pendidikan Matematika, 10*(1), 15–27.
81. Olausson, D., & Berggren, C. (2010). Managing uncertain, complex product development in high tech firms: In search of controlled flexibility. *R&D Management, 40*(4), 383–399.
82. O'Lear, S. (2016). Climate science and slow violence: A view from political geography and STS on mobilizing technoscientific ontologies of climate change. *Political Geography, 52*, 4–13.
83. Osborne, T., & Rose, N. (1999). Do the social sciences create phenomena? The example of public opinion research. *The British Journal of Sociology, 50*(3), 367–396.
84. Page, B. I., Shapiro, R. Y., & Dempsey, G. R. (1987). What moves public opinion? *American Political Science Review, 81*(1), 23–43.
85. Pindyck, R. S. (2013). Climate change policy: What do the models tell us? *Journal of Economic Literature, 51*(3), 860–872.
86. Raleigh, C., & Urdal, H. (2007). Climate change, environmental degradation and armed conflict. *Political Geography, 26*(6), 674–694.
87. Reichlin, P., & Siconolfi, P. (1996). The role of social security in an economy with asymmetric information and financial intermediaries. *Journal of Public Economics, 60*(2), 153–175.
88. Ripple, W. J., Smith, P., Haberl, H., Montzka, S. A., McAlpine, C., & Boucher, D. H. (2014). Ruminants, climate change and climate policy. *Nature Climate Change, 4*(1), 2–5.
89. Roberts, M. R. (2015). The role of dynamic renegotiation and asymmetric information in financial contracting. *Journal of Financial Economics, 116*(1), 61–81.
90. Rodríguez-García, C., García-Pintos, A., Caballero, G., & Vázquez, X. H. (2022). The role of knowledge maps in sub-national climate change policymaking and governance. *Climate Policy, 22*(3), 273–284.

91. Rokeach, M. (1968). The role of values in public opinion research. *Public Opinion Quarterly, 32*(4), 547–559.
92. Rose, N., & Miller, P. (2010). Political power beyond the State: Problematics of government. *The British Journal of Sociology, 61*, 271–303.
93. Rosendorff, B. P., & Milner, H. V. (2001). The optimal design of international trade institutions: Uncertainty and escape. *International Organization, 55*(4), 829–857.
94. Runge, C. F. (1984). Institutions and the free rider: The assurance problem in collective action. *The Journal of Politics, 46*(1), 154–181.
95. Saparwadi, L. (2022). Kesalahan siswa SMP dalam memahami masalah matematika bentuk soal cerita. *MATHEMA: Jurnal Pendidikan Matematika, 4*(1), 1–12.
96. Schäffer, A. A. (1991). Simple local search problems that are hard to solve. *SIAM Journal on Computing, 20*(1), 56–87.
97. Scheffran, J., & Battaglini, A. (2011). Climate and conflicts: The security risks of global warming. *Regional Environmental Change, 11*, 27–39.
98. Schmidt, J., Helme, N., Lee, J., & Houdashelt, M. (2008). Sector-based approach to the post-2012 climate change policy architecture. *Climate Policy, 8*(5), 494–515.
99. Schneider, C. J. (2020). Public commitments as signals of responsiveness in the European Union. *The Journal of Politics, 82*(1), 329–344.
100. Scott, D., & Becken, S. (2010). Adapting to climate change and climate policy: Progress, problems, and potentials. *Journal of Sustainable Tourism, 18*(3), 283–295.
101. Scott, W. A., Rhodes, E., & Hoicka, C. (2023). Multi-level climate governance: Examining impacts and interactions between national and sub-national emissions mitigation policy mixes in Canada. *Climate Policy, 23*(8), 1004–1018.
102. Shove, E. (2010). Beyond the ABC: Climate change policy and theories of social change. *Environment and Planning A, 42*(6), 1273–1285.
103. Simms, C., & Frishammar, J. (2024). Technology transfer challenges in asymmetric alliances between high-technology and low-technology firms. *Research Policy, 53*(3), 104937.
104. Srinivasan, A., Wu, R., Heer, P., & Sansavini, G. (2023). Impact of forecast uncertainty and electricity markets on the flexibility provision and economic performance of highly-decarbonized multi-energy systems. *Applied Energy, 338*, 120825.
105. Thompson, A. (2010). Rational design in motion: Uncertainty and

flexibility in the global climate regime. *European Journal of International Relations, 16*(2), 269–296.

106. Tobondo, Y. A., & Tondowala, S. F. H. (2024). Integrasi psikologi pendidikan dan analisis kesalahan bahasa dalam peningkatan kualitas pembelajaran. *Interdisciplinary Journal (IDe), 2*(1), 57–63.
107. Tsaousoglou, G., Mitropoulou, K., Steriotis, K., Paterakis, N. G., Pinson, P., & Varvarigos, E. (2021). Managing distributed flexibility under uncertainty by combining deep learning with duality. *IEEE Transactions on Sustainable Energy, 12*(4), 2195–2204.
108. Ulfah, U., Prabawanto, S., & Jupri, A. (2017). Students' mathematical creative thinking through problem posing learning. In *Journal of Physics: Conference Series, 895*(11), 012097.
109. Valenzuela, J. M. (2014). Climate change agenda at subnational level in Mexico: Policy coordination or policy competition? *Environmental Policy and Governance, 24*(3), 188–203.
110. Van Ackooij, W., Danti Lopez, I., Frangioni, A., Lacalandra, F., & Tahanan, M. (2018). Large-scale unit commitment under uncertainty: An updated literature survey. *Annals of Operations Research, 271*(1), 11–85.
111. Verdu, A. J., Tamayo, I., & Ruiz-Moreno, A. (2012). The moderating effect of environmental uncertainty on the relationship between real options and technological innovation in high-tech firms. *Technovation, 32*(9, 10), 579–590.
112. Vestby, J. (2019). Climate variability and individual motivations for participating in political violence. *Global Environmental Change, 56*, 114–123.
113. Von Uexkull, N., & Buhaug, H. (2021). Security implications of climate change: A decade of scientific progress. *Journal of Peace Research, 58*(1), 3–17.
114. Walters, W. (2006). Rethinking borders beyond the state. *Comparative European Politics, 4*, 141–159.
115. Watson, R. T. (2005). Turning science into policy: Challenges and experiences from the science–policy interface. *Philosophical Transactions of the Royal Society B: Biological Sciences, 360*(1454), 471–477.
116. Weber, C. L., & Peters, G. P. (2009). Climate change policy and international trade: Policy considerations in the US. *Energy Policy, 37*(2), 432–440.
117. Xu, J. (2021). Conflicts in multi-level governance: An analysis of international climate policy implementation at the sub-national level. *Global Public Policy and Governance, 1*(4), 401–420.

118. Yang, C., Sun, W., Han, D., & Yin, X. (2022). Research on power system flexibility considering uncertainties. *Frontiers in Energy Research, 10*, 967220.

119. Yang, F., Choi, Y. J., Misch, A., Yang, X., & Dunham, Y. (2018). In defense of the commons: Young children negatively evaluate and sanction free riders. *Psychological Science, 29*(10), 1598–1611.

120. Yu, K., Cadeaux, J., Luo, B. N., & Qian, C. (2023). Process ambidexterity driven by environmental uncertainty: Balancing flexibility and routine. *International Journal of Operations & Production Management, 43*(12), 1976–2007.

Chapter 3

AREA-BASED CONSERVATION IN THE TWENTY-FIRST CENTURY

Contents

INTRODUCTION

Governments, policy-makers and many members of the conservation community have long held that protected areas are a fundamental cornerstone of biodiversity conservation. There is a growing recognition of the significance of other effective area-based conservation measures (OECMs). In 2018, the Convention on Biological Diversity (CBD) defined OECMs as areas contributing to effective biodiversity conservation, including certain private conservation initiatives, territories conserved by local communities and Indigenous peoples, and government-managed water catchment areas. Area-based conservation measures, including protected areas and OECMs, are recognized in the 2030 Agenda for Sustainable Development and the CBD. Particularly, the 10-year Strategic Plan for Biodiversity of the CBD—agreed upon by 168 countries in 2010—set a clear goal (Aichi Target 11) to conserve a specific percentage of terrestrial and inland water areas and marine and coastal areas. This target aims to protect areas of significant importance for ecosystem and biodiversity services through well-managed and interconnected systems of protected areas and OECMs, integrated into the wider seascape and landscape by 2020. This goal has been the primary focus of the area-based conservation plans for recent years (Maxwell et al., 2020).

From 2010 to 2019, the percentage of the world's terrestrial and freshwater ecosystems (except Antarctica) covered by protected areas increased from 14.1% to 15.3%, and the percentage of the marine realm from 2.9% to 7.5%. Since 2010, OECMs have emerged as area-based conservation systems, although their global extent cannot be tracked systematically. Nevertheless, despite these promising initiatives, certain troubling spatial patterns within the worldwide protected-area network are becoming increasingly evident. According to a recent analysis, a significant amount of sea and land, approximately 1.1 million km^2, was consistently taken out of global protected-area estates every year from 2006 to 2018 (Cook et al., 2024). It is also worrisome that nations are giving less consideration to the qualitative aspects of Aichi Target 11, such as the importance of having connected and representative protected areas governed and managed fairly. In addition, ongoing challenges such as inadequate funding and ineffective management hinder protected areas to preserve ecosystems and biodiversity effectively. There is a potential danger that humanity may not achieve the overarching strategic objective of enhancing biodiversity by protecting genetic diversity, species, and ecosystems (Cunningham et al., 2021).

The 15th conference of the parties to the CBD will agree on a post-2020 global biodiversity framework. This strategic plan is of utmost importance in addressing the urgent need to prevent a devastating worldwide biodiversity loss. The pressing need for action has strengthened the arguments for a significant global extension of area-based conservation and a thorough overhaul of the formulation and application of environmental targets. Given the current state of global biodiversity, evaluating the successes and shortcomings of area-based conservation initiatives from the past decade is important. This assessment will provide valuable insights into the broader context of the issue (Tracewski et al., 2016).

3.1. THE PERFORMANCE OF PROTECTED AREAS SINCE 2010

Here is a comprehensive analysis of the impact of global protected area expansion on the Aichi Target 11 qualitative components. The analysis covers the period from 2010 to 2019 and includes detailed information on the calculations and methodology used, which can be found in the Supplementary Methods. This section does not mention OECMs because there is currently no global database showing the extent of these sites (Barber et al., 2012) (Figure 3.1).

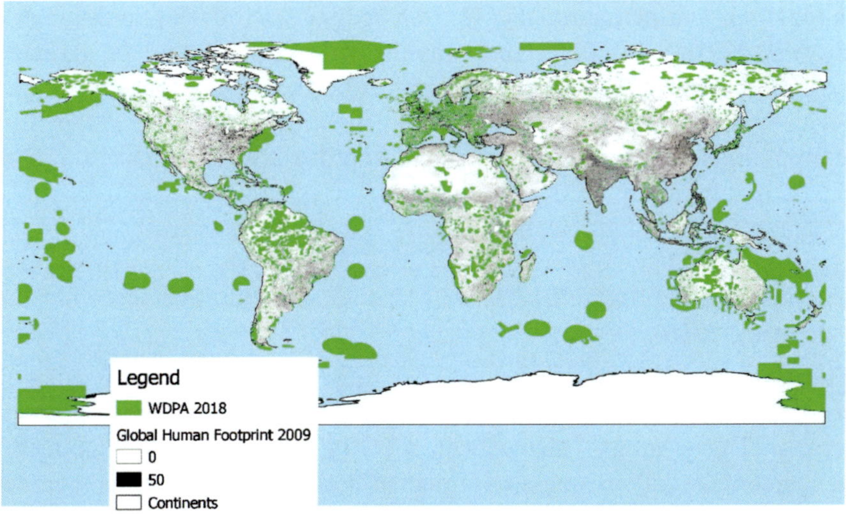

Figure 3.1. Largest protected areas of the world across marine and terrestrial ecosystems in relation to areas of human land use transformation, the Global Human Footprint. WDPA 2014, Accessed from Protected Planet October 2018. Global terrestrial Human Footprint maps for 2009 from Dryad Digital Repository; HF values stretched from 0 (min) to 50 (max).

Source: Baldwin, R.F.; Beazley, K.F. Emerging Paradigms for Biodiversity and Protected Areas. Land 2019, 8, 43. https://doi.org/10.3390/land8030043, Creative Commons License.

3.1.1. Protected Areas Are Ecologically Representative

Ecological representation is often understood as including species or ecoregions, particularly those at risk of extinction, to ensure comprehensive coverage. Our analysis focused on the impact of global protected-area expansion from 2010 to 2019 on the coverage of 12,056 species classified as 'Critically Endangered,' 'Endangered,' or 'Vulnerable' on the IUCN Red List. From 2010 to 2019, there was a slight increase in the percentage of species with an area of its geographical range protected, rising from 86% to 87.6% (n = 10,563). Nevertheless, the proportion of species assessed with sufficient representation within protected areas in 2019 was 21.7% (n = 2,618), an improvement from 18.9% in 2010. Adequacy targets for each species were determined based on their geographical range (Hazen and Anthamatten, 2004).

Over the past decade, there has been a significant increase in the percentage of threatened reef-forming corals with sufficient representation, growing from 9.1% to 44.0%. Over this period, there was an increase in the proportion of species with sufficient coverage for cartilaginous fishes (32.4%), marine bony fishes (42.1%), marine mammals (43.2%), seagrasses (50.0%), and threatened mangroves (50.0%). In 2019, there was insufficient protection for threatened marine reptiles. In the past decade, there has been a growth in the percentage of species with adequate coverage on land. Birds increased from 3% to 33.6%, while amphibians grew by 2% to 10.9%. Reptiles also experienced a growth of 13.6%, mammals by 37.0%, and freshwater species by 19.0%. Notably, a significant majority of threatened species assessed, specifically 78.3% (n = 9,438), lacked sufficient protection as of 2019. Shockingly, a substantial portion of these species, approximately 1,493 (12.4%), had no coverage whatsoever (Watson et al., 2016).

The researchers also evaluated the degree to which the adequacy targets—10% for pelagic regions or marine ecoregions and 17% protection for terrestrial ecoregions —have been met. In 2019, there was an increase in the percentage of adequately protected terrestrial ecoregions compared to 2010. Specifically, 42.6% (n = 361) of these ecoregions were adequately protected. A significant portion (20.6%) of the land that has been protected since 2010 encompasses subtropical and tropical grassland ecoregions, which are considered critically endangered biomes. On the other hand, a significant portion of the land protected since 2010 consisted of desert or dry ecoregions, known for having fewer species, and are already adequately represented in protected areas worldwide (Belote et al., 2017). Over the past decade, there has been a significant increase in the proportion of marine ecoregions with sufficient coverage. This growth is particularly notable around Antarctica in the Southern Ocean, which now accounts for 4.2% of all new marine protected areas, covering an area of 0.7 million km^2. In 2010, the coverage was only 31.8%, but it has now risen to 45.7% (n = 106). Although significant progress has been made in establishing protected areas since 2010, the pelagic region coverage is still lacking. In 2019, only a small percent of these regions, specifically 10.8%, were adequately protected, an improvement from the 2.7% recorded in 2010. In the past decade, a significant amount of new marine protected area, totaling 6.0 million km^2, was focused on the waters between South America and Australia. This region received 36.6% of all the growth of protected areas in pelagic regions (Rice and Houston, 2011).

3.1.2. Coverage of Areas Crucial for Biodiversity

The KBA approach provides a worldwide standard for identifying freshwater, terrestrial, and marine sites that significantly preserve global biodiversity (Marchese, 2015). More than 15,000 KBAs have been identified, with 83.1% being 'Important Bird Areas' specifically focused on avian conservation. Host nations must prioritize managing these sites to safeguard biodiversity, even if they are not designated as protected areas. According to the researchers' analysis, the average coverage of terrestrial KBAs increased from 43.6% in 2010 to 45.9% in 2019. Similarly, the average coverage of marine KBAs increased from 37.9% in 2010 to 43.3% (Figure 3.3). In 2019, a significant number of KBAs, approximately 4,900 (33.0%), still lacked protected area coverage (Hoffmann, 2022).

Wilderness areas are pristine seas and landscapes that remain untouched by human activities, preserving their ecological integrity. They are essential for sustaining planetary life-support systems and are considered particularly crucial for the survival of endangered species, especially in the face of climate change. It is worth noting that more than half (55.6%) of wilderness areas coincide with the geographic range of the minimum of one threatened species. However, it is interesting to observe that wilderness areas are not well aligned with Key Biodiversity Areas (KBAs), as only 1.2% of all land and sea on Earth is recognized as both a wilderness area and a KBA. According to our analysis, there has been a noticeable increase in coverage for both terrestrial and marine wilderness areas over the past decade. Specifically, the coverage for terrestrial areas has risen from 19.7% to 22.1%, while the coverage for marine areas has significantly increased from 2.0% to 8.5% (Vihervaara et al., 2017).

3.1.3. Coverage of Ecosystem Services

The storage and sequestration of carbon in terrestrial ecosystems is a crucial factor in mitigating climate change caused by humans. Consequently, we evaluated the extent of soil carbon stocks and global aboveground biomass.

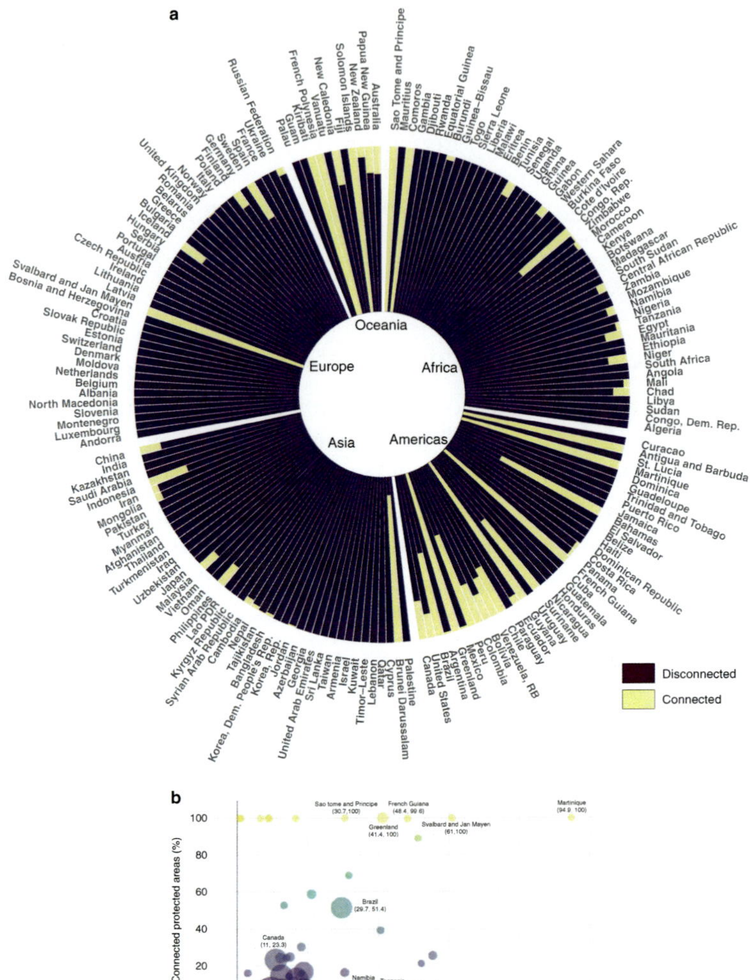

Figure 3.2. *(a) proportion of connected and disconnected PAs for each nation.*

Source: Ward, M., Saura, S., Williams, B. et al. (2020). Just ten percent of the global terrestrial protected area network is structurally connected via intact land. Nat Commun., 11, 4563. https://doi.org/10.1038/s41467-020-18457-x, Creative Commons License.

Countries and territories are grouped per continent and are ordered based on smallest to largest landmass (moving clockwise); (b) Relationship between the proportion of land under protection (x-axis) and the proportion of connected PAs (y-axis) per country and territory (less connected countries and territories are purple while more connected countries and territories are yellow). The size of the bubble indicates the size of the country and territory.

Between 2010 and 2019, the aboveground biomass cover rose from 22.6% to 23.7%. The average coverage worldwide of soil-carbon stocks was less and experienced a smaller increase over the past decade. In 2010, it stood at 13.9%; in 2019, it reached 14.6% (400.5 Pg C). Significant amounts of soil carbon remain unprotected in various regions, including Southeast Asia, Russia, and Northeast North America (Lemieux et al., 2011).

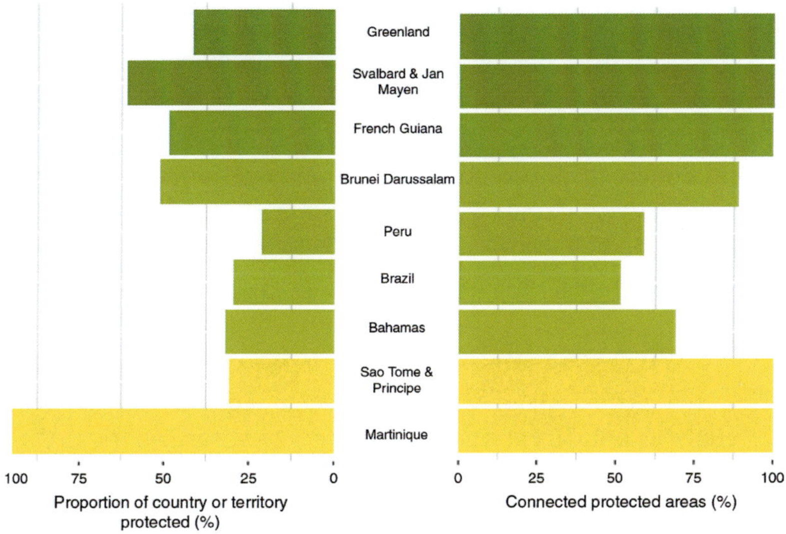

Figure 3.3. Only nine countries and territories protect >17% of land (left) and maintain >50% connected PAs (right). Countries and territories are ordered based on the highest proportion of remaining intact land (less intact countries and territories are yellow, while more intact countries and territories are green).

Source: Ward, M., Saura, S., Williams, B. et al. (2020) Just ten percent of the global terrestrial protected area network is structurally connected via intact land. Nat. Commun., 11, 4563. https://doi.org/10.1038/s41467-020-18457-x, Creative Commons License.

Nevertheless, ongoing improvements in mapping terrestrial organic carbon, especially in tropical rainforests and peatlands, can shape the future knowledge of carbon storage in these specific regions. The "biological pump," made up of phytoplankton in the seas moving carbon to the deep ocean, is also very important for preventing climate change because it takes carbon out of the oceans and atmosphere for thousands of years. It is estimated that within marine protected areas, 0.17 Pg of dissolved organic carbon and 0.21 Pg of particulate organic carbon are exported annually. Nevertheless, the factors influencing carbon export in the oceans exhibit seasonal variations, and the effectiveness of such areas in promoting carbon export may fluctuate over time (Geldmann et al., 2015) (Figure 3.2).

With three billion people depending on farmed or wild-caught seafood as their main source of protein, ensuring a consistent seafood supply is a crucial global objective. The coverage of the least-productive exclusive economic zones, or those in the bottom 20% of annual fisheries catch per unit area, was, on average, three times greater than the coverage of the most-productive exclusive economic zones, or those in the top 20% of annual fisheries catch per unit area (31.2% versus 9.8%) when the protected area coverage of the least- and most-productive marine regions for fisheries catch in the oceans (Morales et al., 2015).

In addition, there has been no significant change in the average coverage of the most productive EEZ since 2010. Moreover, it was discovered that there is no official protected area coverage in the seven pelagic regions that are most productive for fisheries capture.

3.1.4. Protected Areas Being Well-Connected

Recognizing the significance of interconnected ecosystems is crucial for preserving vital evolutionary and ecological processes, particularly in the face of rapid environmental and climate shifts. Enhancing connectivity among marine protected areas is crucial for replenishing and maintaining fish populations, even on fished reefs. According to a previous study, the coverage of protected areas inside the possible dispersal distance of terrestrial vertebrates was quite low in 2016, with only 30% of terrestrial ecoregions meeting the minimum requirement of 17% coverage. Another study revealed that between 2010 and 2018, there was an increase in the proportion of linked terrestrial protected areas from 6.5% to 7.7%. Unfortunately, the assessments failed to consider the permeability of unprotected land between the protected areas (Santini et al., 2016). Studies conducted at the regional level reveal little connectivity between freshwater and marine protected areas, particularly for species with a dispersive larval stage. However, assessments of connectivity at the global level have not been conducted (Belote et al., 2016).

3.1.5. Effectiveness of Protected Area Management

Four major methodologies have been applied in the last few decades to assess area-based conservation initiatives (Table 3.1).

Table 3.1. *Description of the Main Threats in Protected Areas*

Threats	Description
Climate change	Climate change refers to the effect on the distribution, abundance, and quality of species or ecosystems and increasing threat from fires due to global warming.
Deforestation	Logging of timber species for commercial purposes within and around PAs.
Engineering Work (Infrastructure and roads)	Construction of new infrastructures linked to the different activities, mainly oil and mining. Additionally, the opening of roads for commercial purposes.
Fires	Direct impact on species, ecosystems, or cultural resources due to natural or anthropogenic fires.
Poaching	Activity that consists of killing an individual of a species for entertainment or commercial purposes.
Invasive Alien Species	Direct (elimination) or indirect (competition; disease transmission) impact on species or ecosystems due to the presence of exotic species or species introduced by anthropogenic activities such as tourism, livestock, trade, transportation, among others, within PAs.
Land use change	Direct (destruction) or indirect (isolation) impact on species or ecosystems due to changes in land use.
Aggressive tourism activities	Activities of any of the tourism segments developed within the SNAP.
Wildlife trafficking	Direct or indirect impact on species, ecosystems, or natural resources due to the direct action of complete or partial extraction of flora, fauna, or biological resources of any kind.

Source: Mestanza-Ramón C, Monar-Nuñez J, Guala-Alulema P, Montenegro-Zambrano Y, Herrera-Chávez R, Milanes CB, Arguello-Guadalupe C, Buñay-Guisñan P, Toledo-Villacís M. (2023). A review to update the protected areas in ecuador and an analysis of their main impacts and conservation strategies. Environments, 10(5):79. https://doi.org/10.3390/environments10050079.

Three of these approaches focus on management effectiveness, with the first one being input evaluation, which assesses the management resources sufficiency for area-based conservation. According to a recent study, 47% of protected places on land suffer from a lack of staff and funding, with the Neotropics having particularly low levels of resource availability. This study covered around 23% of all terrestrial protected areas. Similarly, a comprehensive analysis of 433 marine protected areas revealed that a staggering 65% of them lack the necessary funding to meet their basic management requirements. Furthermore, 91% of these areas suffer from a shortage of on-site staff, with their capacity being either inadequate or below the optimum level. Other factors, such as the lack of strict enforcement of regulations in protected areas, have also been linked to ineffective management (Kolahi et al., 2013).

Another evaluation approach, known as threat reduction evaluation, examines the effectiveness of area-based conservation in reducing threats to biodiversity. Most of these evaluations indicate that protected areas effectively reduce human stresses within their borders, although they may not eliminate them. As a seasoned environmental policy expert, it is worth noting that the human pressures within 55% of land-based protected areas escalated from 1993 to 2009. Similarly, 94% of marine protected areas established before 2014 allow fishing activities (Stoll-Kleemann, 2010). Through expertise in environmental policy, researchers have discovered that terrestrial protected areas have proven effective in reducing forest degradation and deforestation, surpassing the results seen in nearby unprotected areas such as the Amazon. Additionally, marine protected areas can decrease fishing vessel traffic and mitigate the harmful impacts of certain non-native species (Lemieux et al., 2019).

One approach to evaluation, known as outcome evaluation, examines whether area-based conservation goals are being met compared to a scenario with no intervention. A controlled study found that approximately 12,000 protected areas failed to effectively reduce human pressures within their borders from 1995 to 2010. Nevertheless, numerous studies have documented the positive impacts of protected areas on biodiversity. For instance, a comprehensive study of 359 terrestrial protected areas revealed that abundance was 14.5% higher and species richness was 10.6% higher within protected areas than outside. These effects were particularly significant in human-dominated land uses in tropical regions, such as plantations and cropland. A study conducted on 218 marine protected areas discovered that the fish biomass inside these protected areas is almost twice as much as in non-protected sites (Ervin, 2003). This finding highlights the positive impact of implementing protective measures for marine ecosystems.

Marine protected areas may contribute to the restoration of commercial fish species. Specifically, no-take marine reserves are particularly effective at restoring trophic function, lowering levels of coral disease, and increasing species diversity, density, and biomass in temperate and tropical systems. Finally, numerous studies have examined the social impacts of protected areas. For instance, a comprehensive study conducted on 603 protected areas revealed that households residing near protected areas with tourism prospects experienced a 16% decrease in the likelihood of poverty and a 17% increase in wealth levels compared to similar households residing far from protected areas (Hockings et al., 2015).

3.1.6. Equitable Governance and Management in Protected Areas

Considering the various aspects of social equity within protected areas is crucial. This includes ensuring fair distribution of burdens and benefits, promoting participatory and transparent decision-making, and respecting traditional management practices, customary rights, and cultural identities. According to a recent survey that covered 225 protected areas, most community representatives, staff, and conservation managers believe that the advantages of protected areas are distributed fairly. However, the study revealed an inequitable decision-making process in many instances. Additionally, local stakeholders expressed concerns about a perceived loss of rights over natural resources following the establishment of a protected area (Zafra et al., 2017).

Protected areas that include local populations as stakeholders frequently produce superior conservation and socio-economic benefits despite the paucity of evidence supporting progress toward social fairness. Stakeholder participation, surveillance, political will, leadership, and the presence of conflict resolution and sanctioning procedures were determined to be critical to accomplishing ecological goals in an evaluation of 27 maritime protected areas. Long-standing, no-take marine protected areas demonstrate successful conservation and favorable governance and economic results for reliant human communities. In addition, it has been observed that community-managed terrestrial protected areas have shown greater effectiveness in reducing deforestation pressures compared to nationally designated protected areas. This has been observed in various countries, such as Namibia, Australia, Brazil, and Peru (Campbell and Gray, 2019).

3.2. LESSONS LEARNED AND PRIORITY ACTIONS

In the past decade, national governments have made some progress toward Aichi Target 11, especially in the marine realm. Nevertheless, countries have not yet achieved this goal. To achieve 17% coverage for freshwater and land environments, the expansion rate of terrestrial protected areas should have been twice as high as observed in the past decade. In 2019, a significant number of threatened species and ecosystems and threatened species on sea and land still lacked sufficient protection, with more than half of all ecosystems and 78.3% of identified threatened species falling into this category. It is evident from this assessment that nations need to enhance their area-based conservation efforts to ensure that they make a meaningful contribution to global goals for ecosystem and species conservation (Rhodes and Dawson, 2013). These goals encompass everything from preventing extinction to preserving the integrity of ecosystems.

Over the past decade, it has become evident that numerous protected areas suffer from inadequate management, primarily due to chronic resource shortages. Additionally, there is a concerning lack of proper representation for community and indigenous groups in conservation plans for sea, water, and land. Three pressing challenges have been identified that require immediate attention from policy-makers, scientists, governments, and other stakeholders as they enter the subsequent decade of area-based conservation (Kinney et al., 2015) (Table 3.2).

Table 3.2. Main Hypotheses Relating Effects of Marine Protected Areas (MPA) on Site (alpha) Stability, Species Stability and Asynchrony, Compared to Open Areas (OA)

Hypothesized Relationships	**Predicted Effect**	**Mechanisms**
Within sites (communities)		
MPA → alpha stability	MPA are more stable than OA.	Greater stability (lower fluctuations) in abundance of individual species and greater functional richness increases stability in MPA compared to OA.
MPA → species stability → alpha stability	Stronger positive relationship between alpha stability and species stability in MPA than OA.	MPA increase alpha stability by maintaining more stable populations compared to OA.
§MPA → asynchrony → alpha stability	The contribution of asynchrony to stability is stronger (weaker) in MPA than OA.	MPA increase (decrease) asynchrony if relieve from fishing and direct human disturbances induce divergent (coherent) temporal fluctuations in fishes.
MPA → MHW → species and alpha stability	Alpha and species stability decline more abruptly with intensifying MHW in OA than MPAs.	MPA buffer reef fishes from MHW by maintaining greater population abundances, functional richness and asynchronous fluctuations, all of which contribute to increase stability.

MPA → Remoteness → species and alpha stability	Stronger positive relationship between alpha and species stability with remoteness in MPA than OA.	Remote MPA are relieved from both fishing and direct human impacts, whereas fishing can still impact remote sites in OA, decreasing alpha and species stability.
MPA → temperature niche	Stronger positive relationship between the abundance of thermally resistant species with intensifying MHW in MPA than OA.	MPA support higher fish abundances and promote stability by allowing thermally resistant species to attain large population sizes under intensifying warming conditions.
Among sites (metacommunities)		
MPA → gamma stability and underlying mechanisms	Greater asynchrony and stability in MPA than OA at the metacommunity scale.	Local effects of MPA on species stability and asynchrony scale-up at the metacommunity level.
Size of MPA network → gamma stability and underlying mechanisms	Differences between MPA and OA increase with the size of MPA networks.	Gamma stability and the underlying mechanisms operate more strongly in large MPA networks embracing a wide range of habitats and environmental conditions.

1. MPAs are hypothesized to mediate effects of Marine Heatwaves (MHW) and remoteness (distance from large cities) at local scales. At the metacommunity scale, gamma stability is related to multiple underlying mechanisms, including population and species stability and population, species and spatial asynchrony (Figure 3. 1).

2. §There is no clear a priori expectation about the direction of the effect of MPAs on asynchrony and both positive and negative effects are considered here.

Source: Benedetti-Cecchi, L., Bates, A.E., Strona, G. et al. (2024). Marine protected areas promote stability of reef fish communities under climate warming. Nat. Commun., 15, 1822. https://doi.org/10.1038/s41467-024-44976-y.

3.2.1. Making OECMs Count

Increasing opportunities exist to officially acknowledge areas beyond government-managed protected areas, potentially preserving biodiversity. Furthermore, there is a growing recognition of other OECMs and protected areas governed indigenously and privately. It wasn't until November 2018 that the criteria and guiding principles for identifying OECMs were finally agreed upon, even though their significance was acknowledged in Aichi Target 11 in 2010. This delay has likely played a role in the lack of attention given to OECMs in national biodiversity strategies and policies in recent years. Having reached a formal agreement on the definition, managing bodies and nations are poised to expedite the implementation of OECMs. The conservation community must ensure that OECMs significantly contribute to biodiversity conservation (Lemieux et al., 2022).

One potential solution to the lack of representation in the global protected-area estate is the implementation of OECMs. A recent study has revealed that potential OECMs partially cover 566 unprotected KBAs. Interestingly, compared to nationally designated protected areas, OECMs may be acceptable to society in productive sea- and landscapes, which are known for being areas for poorly protected threatened species. Understanding the importance of OECMs in coastal marine habitats, farmlands, and managed forests can greatly improve the effectiveness of area-based conservation initiatives. Maintaining and restoring the natural ecological functions in these areas is crucial to enhance connectivity. Greater acknowledgment of OECMs is expected to contribute to a more balanced approach to area-based conservation management, ensuring that it is inclusive and benefits a wide range of stakeholders. It is evident from a recent study that the lands under the stewardship of Indigenous communities in Canada, Brazil, and Australia harbor a rich diversity of vertebrate species comparable to those found within officially designated protected areas. This finding underscores the significance of collaborating with Indigenous peoples to acknowledge and respect OECMs within their territories (Fitzsimons et al., 2024).

To fully realize the benefits of OECMs, the conservation community, private industry, and governments must rally behind them swiftly. This collective effort is necessary to address numerous protected areas' challenges, such as insufficient resourcing and reporting. In December 2019, a reporting platform for OECMs was launched, which can significantly enhance the accuracy of progress assessments toward the follow-up of Aichi Target 11, provided that countries actively utilize it. For OECMs to be successful, conservation actors and governments must prioritize social safeguards and human rights, especially in community and indigenous areas (Laffoley et al., 2017). When considering the need for adjustments to livelihoods to meet OECM criteria, it is crucial to carefully create substitute livelihood opportunities that provide tangible advantages to resource users. It is important for alternative livelihood schemes to consider the preservation of biodiversity in OECMs (Fitzsimons et al., 2024).

3.2.2. Tracking the Increasing Dynamism of Area-Based Conservation

Recent studies have revealed that protected areas exhibit greater temporal and spatial dynamics than previously believed. It is quite challenging to evaluate the vulnerability of ecosystems or the impact of changes on area-based conservation networks due to the lack of proper documentation regarding decisions to relax, shrink, or remove protected areas. Transparency in reporting protected area dynamics is crucial for the conservation community, particularly when they negatively impact biodiversity outcomes (Maxwell et al., 2020).

There have been numerous removals from the protected-area estate due to events known as 'protected area downgrading, downsizing, and degazettement' (PADDD). Between 1997 and 2014, over one-third of Australia's protected area network, covering 416,740 km^2, was impacted by over 1,500 PADDD events. In addition, natural World Heritage sites, including the Yosemite, Serengeti, and Virunga National Parks, have been impacted by 23 PADDD events. Fragmentation and forest loss can be accelerated by PADDD events, with a majority (62%) being linked to activities that directly contradict biodiversity conservation efforts (Hoffmann, 2022).

These activities often involve infrastructure development and industrial-scale resource extraction. There is a significant cause for concern regarding the numerous undocumented PADDD events, especially in private lands and marine systems (Dixon et al., 2016).

The conservation community and governments must increase their engagement with global PADDD tracking platforms, such as www.padddtracker.org, to enhance the openness of area-based conservation decisions. Incorporating PADDD tracking data into current area-based conservation databases, such as the World Database on Protected Areas, could enhance their usefulness and support global reporting (Langhans et al., 2019). There are indications of efforts to expand or improve protected areas via better resources and management or by implementing stricter regulations, which could bring about positive changes in area-based conservation. Therefore, it is also evident that improved incentives and monitoring of the range of modifications to protected areas are needed to enhance their capacity to preserve biodiversity. It is recommended that these changes be referred to as 'protected area gazettement, expansion and enhancement' (PAGEE) events. Ensuring clear and transparent tracking of PAGEE and PADDD events is crucial in addressing and avoiding any further negative impacts on area-based conservation (Crespo et al., 2020) (Table 3.3).

Table 3.3. Addressing Critical Challenges to Enhance Progress Towards Goals for Area-Based Conservation Within the 21st Century

	New Conservation Science Approach	Half Earth Approach	Biocultural Approaches
What?	Biodiversity conservation, especially ecosystem services	Biodiversity conservation, especially species and ecosystem integrity	Multiple objectives of relevant stakeholders
Why?	Emphasis on instrumental value of biodiversity	Emphasis on intrinsic value of biodiversity	Pluralistic worldviews (including relational (7), intrinsic, instrumental values)
Where?	More emphasis on human-dominated landscapes	More emphasis on remote locations and wilderness	Tailor interventions to social-ecological context
How?	Emphasis on economic incentives and payment for ecosystem services	Emphasis on strict protected areas	Diverse and nested institutions with adaptive governance and management
Who?	Not specified	Not specified	Partnerships and social learning among multiple stakeholders

Source: Gavin MC, McCarter J, Berkes F, Mead ATP, Sterling EJ, Tang R, Turner NJ. (2018). Effective biodiversity conservation requires dynamic, pluralistic, partnership-based approaches. Sustainability, 10(6):1846. https://doi.org/10.3390/su10061846.

3.2.3. Outcome-Orientated Evaluation of Area-Based Conservation

Various approaches have been developed to assess area-based conservation efforts, all of which have their merits. However, the conservation community still heavily relies on evaluation methods, primarily focusing on threat reduction or management inputs. It is now very difficult to adopt assessment methods that better capture area-based conservation's socio-economic outcomes and biodiversity (Sverdrup et al., 2014).

The Global Database on Protected Area Management Effectiveness offers comprehensive information on various aspects of protected area processes, such as the effectiveness of law enforcement activities effectiveness or the presence of a management plan. However, most of the methodologies used by the Global Database on Protected Area Management Effectiveness were not designed to measure the impact of protected area management on ecosystems and species. As a result, they cannot be utilized to assess the progress made in effectively conserving biodiversity. Conservation assessments that are more outcome-oriented are made possible by Hi-Res maps of ecological shifts in terrestrial and marine environments, including changes in forest cover and cumulative human pressure. However, local biodiversity patterns are not always solely attributed to sea and landscape ecological changes. Cumulative mapping of human pressure likewise lags behind forest-cover mapping initiatives in terms of temporal resolution, and some human pressure maps have spatial resolutions (77 km^2) that make it impossible to analyze a large number of tiny (<1 km^2) yet critically essential protected areas (Bhola et al., 2021).

Standardizing reporting requirements are proposed for all organizations involved in area-based conservation to include rigorous outcome assessment approaches, both ex-ante to help site areas and ex-post to report on outcomes, to render area-based conservation assessments more outcome-oriented. Conservation agencies must allocate more funding toward comprehensive and carefully planned biodiversity monitoring programs. Monitoring control sites resembling protected areas regarding anthropogenic and ecological conditions will be crucial for these programs. This will help identify and separate any potential confounding effects (Jetz et al., 2022). It is important to promote the involvement of governments and communities in citizen science initiatives that utilize innovative techniques like acoustic monitors, drones, and camera traps. These methods can greatly enhance biodiversity data collection by expanding its coverage and frequency. By leveraging the latest advancements in field campaigns and remote sensing, such as the Biomass Earth Explorer by the European Space Agency, along with in situ reporting of protected area biodiversity trends and management capacity using the Spatial Monitoring and

Reporting Tool (SMART), we can greatly enhance the accuracy and reliability of outcome evaluations (Hong et al., 2017).

3.3. FUTURE-PROOFING AREA-BASED CONSERVATION

While the three mentioned challenges are of utmost importance, it is crucial to implement more comprehensive policy reforms to ensure that area-based conservation can effectively contribute to the long-term objectives set by the CBD. Specifically, the CBD aims to value, conserve, restore, and utilize biodiversity by 2050 while providing essential benefits to all people, sustaining a healthy planet, and maintaining ecosystem services. This last part lists several prerequisites that governments must meet immediately to guarantee the long-term viability of area-based conservation programs. These prerequisites include having enough funding, being climate-smart, and mainstreaming biodiversity throughout national policy frameworks (Mazor et al., 2021).

3.3.1. Secure Adequate Financing

Over the past two decades, there has been a significant decline in global funding for species protection. In the 2000s, the funding stood at around US$200 million per year, but by 2010, it dropped to no more than US$100 million per year. Growing area-based conservation efforts further exacerbate resource shortages at existing sites, leading to increased costs. According to an estimate, the annual cost of managing and protecting a more comprehensive terrestrial protected-area network could amount to US$76.1 billion. Therefore, it is likely that there is currently a budget deficit for area-based conservation that is more than a few billion dollars in size. Addressing this shortfall may prove challenging in the next decade, but private industry and governments must prioritize its reduction immediately (Tambi and Bime, 2019).

If the significant economic contributions of area-based conservation were properly acknowledged, it would be possible to meet future and current resourcing needs. Trips to protected areas generate a significant direct value of US$600 billion annually. Governments must improve their understanding of how area-based conservation efforts contribute to national economies. When establishing a budget for area-based conservation, governments should utilize predictive measures to determine funding requirements and assess the potential effects. Additionally, they need to consider the often-overlooked cost-saving benefits of effective biodiversity conservation (Okpala, 1994). For instance, it would be valuable to compare the expenses resulting from the socio-economic destruction caused by zoonotic diseases like coronavirus disease 2019 or severe acute respiratory syndrome with the costs required to efficiently oversee area-based conservation networks that diminish the supply to illegal wildlife markets. There is a pressing need to enhance the utilization of philanthropic and industry contributions to area-based conservation. It can be achieved by implementing improved funding guidelines that safeguard against any potential compromise to the management or siting of area-based conservation due to private interests (Chapinduka et al., 2007).

3.3.2. Being Climate-Smart

Climate change will play a significant role in determining the effectiveness of area-based conservation efforts in the coming years, as numerous biological responses to climate change are already occurring. According to a recent study, if greenhouse gas emissions continue at their current rate, the average sea-surface temperatures inside marine protected areas will rise by 2.8°C by 2100. Climate change projections were overlaid on the terrestrial protected-area network under a moderate emissions scenario (RCP 4.5). It was discovered that temperatures in the warmest quarter are expected to increase by an average of 2.9°C on protected land by 2050, with European nations experiencing even higher increases.

By 2050, certain biodiverse nations such as Suriname and Guyana may experience a significant decrease of 30–40% in rain on protected land throughout dry months. These shifts in bioclimatic conditions have the potential to significantly disrupt ecological networks and put certain ecosystems and species at risk. By 2040–2050, even the RCP 4.5 scenario will probably result in the disappearance of most warm-water coral reefs (Scherr et al., 2012). There are established methods for integrating climate change in area-based conservation plans. These methods involve restoring or protecting ecosystems surrounding protected areas to maintain ecological connectivity. It is crucial to prioritize equipping managers with the necessary capacity and knowledge to implement practical climate adaptation policies effectively.

The climate resilience of area-based conservation efforts can be enhanced by considering human responses to climate change and enforcing stronger conservation regulations. Caution must be exercised when targeting protected areas in sites where bioclimatic changes may be small (refugia), as the biotic response and predicted changes can vary spatially. For instance, a mere 3.5% of marine protected areas coincide with refugia, providing suitable sea-surface oxygen and temperature concentration. When faced with these situations, utilizing decision-support tools such as systems modeling and value-of-information analysis can help assess the advantages of addressing uncertainties regarding ecological responses to climate change before implementing conservation measures. This approach can result in stronger and more reliable management decisions (Chandra et al., 2018).

3.3.3. Make Biodiversity Conservation Mainstream

Regardless of the quality of the location, resources, or management, area-based conservation can only address a limited number of threats to the long-term survival of biodiversity. Addressing significant, far-reaching challenges necessitates implementing additional measures prompted by comprehensive sea, water, and land management policies. China has taken a significant step by implementing the Ecological Redline Policy, which requires local governments to incorporate ecosystem and biodiversity service assessments into their land-use planning. It is important to consider the potential impact of China's planned activities outside their national boundaries, such as the BRI, on areas of critical conservation concern. The Ecological Redline Policy should be extended to address these concerns. Unfortunately, many national sea, water, and land management policies prioritize economic development over environmental concerns or have loopholes that result in negative environmental consequences. Governments must acknowledge the importance of implementing effective policies, as this will ultimately alleviate the pressure on area-based conservation strategies in the future (Huntley, 2014). Two key changes could greatly enhance the effectiveness of national sea, water, and land management policies. First, it is recommended that nations embrace an ambitious goal for biodiversity, such as achieving a net positive impact on biodiversity. Subsequently, they should establish targets that address environmental and socio-economic aspects, which can effectively contribute to this overarching goal. Targets should be made mutually conditional. Meeting environmental targets, such as protecting 30% of the land, should not be considered successful if the achievement of socio-economic targets, like eliminating incentives harmful to biodiversity, is lacking. Enhancing biodiversity accounting protocols has the potential to greatly improve the effectiveness of sea, water, and land management policies (Karlsson et al., 2017). One example is the shift from the biodiversity impact-offsetting protocols that only move conservation funding around or maintain rates of biodiversity loss (such as averted loss offsetting) to newer protocols that link compensation to the desired progress for endangered species or ecosystems (such as target-based compensation) (Wilson, 2023).

3.4. SUMMARY

Area-based conservation will continue to be crucial for preserving biodiversity well into the 21st century. However, governments have had a significant lack of investment in protected areas, other effective area-based conservation measures (OECMs), and a failure to provide adequate legal protection for these areas. Furthermore, it is crucial for conservation organizations to not only address current shortcomings but also implement evaluation measures that focus on the impact of their efforts and advocate for fair management and government practices. It is crucial for organizations to enhance the accountability of decisions that impact resource and spatial dynamics and to ensure that OECMs can make a significant contribution to biodiversity conservation. To ensure the long-term success of area-based conservation efforts, governments must prioritize securing sufficient funding, adopting climate-smart approaches, and integrating biodiversity considerations into socio-economic and environmental policies.

REFERENCES

1. Barber, C. P., Cochrane, M. A., Souza Jr, C., & Veríssimo, A. (2012). Dynamic performance assessment of protected areas. *Biological Conservation, 149*(1), 6–14.
2. Belote, R. T., Dietz, M. S., Jenkins, C. N., McKinley, P. S., Irwin, G. H., Fullman, T. J., & Aplet, G. H. (2017). Wild, connected, and diverse: Building a more resilient system of protected areas. *Ecological Applications, 27*(4), 1050–1056.
3. Belote, R. T., Dietz, M. S., McRae, B. H., Theobald, D. M., McClure, M. L., Irwin, G. H., & Aplet, G. H. (2016). Identifying corridors among large protected areas in the United States. *PLoS One, 11*(4), e0154223.
4. Bhola, N., Klimmek, H., Kingston, N., Burgess, N. D., van Soesbergen, A., Corrigan, C., & Kok, M. T. (2021). Perspectives on area-based conservation and its meaning for future biodiversity policy. *Conservation Biology, 35*(1), 168–178.
5. Campbell, L. M., & Gray, N. J. (2019). Area expansion versus effective and equitable management in international marine protected areas goals and targets. *Marine Policy, 100*, 192–199.
6. Chandra, A., McNamara, K. E., & Dargusch, P. (2018). Climate-smart agriculture: Perspectives and framings. *Climate Policy, 18*(4), 526–541.
7. Chapinduka Nyasulu, E., & Cloete, C. E. (2007). Lack of affordable housing finance in Malawi's major urban areas. *Property Management, 25*(1), 54–67.
8. Cook, C. N., Lemieux, C. J., & Kroner, R. G. (2024). Increasing transparency and accountability in global protected and conserved area reporting. *One Earth, 7*(3), 421–430.
9. Crespo, G. O., Mossop, J., Dunn, D., Gjerde, K., Hazen, E., Reygondeau, G., & Halpin, P. (2020). Beyond static spatial management: Scientific and legal considerations for dynamic management in the high seas. *Marine Policy, 122*, 104102.
10. Cunningham, C. A., Crick, H. Q., Morecroft, M. D., Thomas, C. D., & Beale, C. M. (2021). Translating area-based conservation pledges into efficient biodiversity protection outcomes. *Communications Biology, 4*(1), 1043.
11. Dixon, M. J. R., Loh, J., Davidson, N. C., Beltrame, C., Freeman, R., & Walpole, M. (2016). Tracking global change in ecosystem area: The Wetland Extent Trends index. *Biological Conservation, 193*, 27–35.

12. Ervin, J. (2003). Rapid assessment of protected area management effectiveness in four countries. *BioScience, 53*(9), 833–841.
13. Fitzsimons, J. A., Partridge, T., & Keen, R. (2024). Other Effective Area-based Conservation Measures (OECMs) in Australia: Key considerations for assessment and implementation. *Conservation, 4*(2), 176–200.
14. Geldmann, J., Coad, L., Barnes, M., Craigie, I. D., Hockings, M., Knights, K., & Burgess, N. D. (2015). Changes in protected area management effectiveness over time: A global analysis. *Biological Conservation, 191*, 692–699.
15. Hazen, H. D., & Anthamatten, P. J. (2004). Representation of ecological regions by protected areas at the global scale. *Physical Geography, 25*(6), 499–512.
16. Hockings, M., Leverington, F., & Cook, C. (2015). Protected area management effectiveness. In *Protected Area Governance and Management* (pp. 889–928).
17. Hoffmann, S. (2022). Challenges and opportunities of area-based conservation in reaching biodiversity and sustainability goals. *Biodiversity and Conservation, 31*(2), 325–352.
18. Hong, J. P., Shim, Y. J., & Heo, H. Y. (2017). Identifying other effective area-based conservation measures for expanding national protected areas. *Journal of the Korean Society of Environmental Restoration Technology, 20*(6), 93–105.
19. Huntley, B. J. (2014). Good news from the South: Biodiversity mainstreaming—A paradigm shift in conservation? *South African Journal of Science, 110*(9, 10), 01–04.
20. Jetz, W., McGowan, J., Rinnan, D. S., Possingham, H. P., Visconti, P., O'Donnell, B., & Londoño-Murcia, M. C. (2022). Include biodiversity representation indicators in area-based conservation targets. *Nature Ecology & Evolution, 6*(2), 123–126.
21. Karlsson-Vinkhuyzen, S., Kok, M. T., Visseren-Hamakers, I. J., & Termeer, C. J. (2017). Mainstreaming biodiversity in economic sectors: An analytical framework. *Biological Conservation, 210*, 145–156.
22. Kinney, M. V., Cocoman, O., Dickson, K. E., Daelmans, B., Zaka, N., Rhoda, N. R., & Darmstadt, G. L. (2015). Implementation of the every newborn action plan: Progress and lessons learned. In *Seminars in Perinatology, 39*(5), 326–337.
23. Kolahi, M., Sakai, T., Moriya, K., Makhdoum, M. F., & Koyama, L. (2013). Assessment of the effectiveness of protected areas management

in Iran: Case study in Khojir National Park. *Environmental Management, 52*, 514–530.

24. Laffoley, D., Dudley, N., Jonas, H., MacKinnon, D., MacKinnon, K., Hockings, M., & Woodley, S. (2017). An introduction to 'other effective area-based conservation measures' under Aichi Target 11 of the Convention on Biological Diversity: Origin, interpretation and emerging ocean issues. *Aquatic Conservation: Marine and Freshwater Ecosystems, 27*, 130–137.
25. Langhans, S. D., Domisch, S., Balbi, S., Delacámara, G., Hermoso, V., Kuemmerlen, M., & Jähnig, S. C. (2019). Combining eight research areas to foster the uptake of ecosystem-based management in fresh waters. *Aquatic Conservation: Marine and Freshwater Ecosystems, 29*(7), 1161–1173.
26. Lemieux, C. J., Beechey, T. J., & Gray, P. A. (2011). Prospects for Canada's protected areas in an era of rapid climate change. *Land Use Policy, 28*(4), 928–941.
27. Lemieux, C. J., Gray, P. A., Devillers, R., Wright, P. A., Dearden, P., Halpenny, E. A., & Beazley, K. (2019). How the race to achieve Aichi Target 11 could jeopardize the effective conservation of biodiversity in Canada and beyond. *Marine Policy, 99*, 312–323.
28. Lemieux, C. J., Kraus, D. T., & Beazley, K. F. (2022). Running to stand still: The application of substandard OECMs in national and provincial policy in Canada. *Biological Conservation, 275*, 109780.
29. Marchese, C. (2015). Biodiversity hotspots: A shortcut for a more complicated concept. *Global Ecology and Conservation, 3*, 297–309.
30. Maxwell, S. L., Cazalis, V., Dudley, N., Hoffmann, M., Rodrigues, A. S., Stolton, S., & Watson, J. E. (2020). Area-based conservation in the twenty-first century. *Nature, 586*(7828), 217–227.
31. Mazor, T., Runting, R. K., Saunders, M. I., Huang, D., Friess, D. A., Nguyen, N. T., & Lovelock, C. E. (2021). Future-proofing conservation priorities for sea level rise in coastal urban ecosystems. *Biological Conservation, 260*, 109190.
32. Morales-Hidalgo, D., Oswalt, S. N., & Somanathan, E. (2015). Status and trends in global primary forest, protected areas, and areas designated for conservation of biodiversity from the Global Forest Resources Assessment 2015. *Forest Ecology and Management, 352*, 68–77.
33. Okpala, D. C. (1994). Financing housing in developing countries: A review of the pitfalls and potentials in the development of formal housing finance systems. *Urban Studies, 31*(9), 1571–1586.

34. Rhodes, L., & Dawson, R. (2013). Lessons learned from lessons learned. *Knowledge and Process Management, 20*(3), 154–160.
35. Rice, J., & Houston, K. (2011). Representativity and networks of marine protected areas. *Aquatic Conservation: Marine and Freshwater Ecosystems, 21*(7), 649–657.
36. Santini, L., Saura, S., & Rondinini, C. (2016). Connectivity of the global network of protected areas. *Diversity and Distributions, 22*(2), 199–211.
37. Scherr, S. J., Shames, S., & Friedman, R. (2012). From climate-smart agriculture to climate-smart landscapes. *Agriculture & Food Security, 1*, 1–15.
38. Stoll-Kleemann, S. (2010). Evaluation of management effectiveness in protected areas: Methodologies and results. *Basic and Applied Ecology, 11*(5), 377–382.
39. Sverdrup-Thygeson, A., Bendiksen, E., Birkemoe, T., & Larsson, K. H. (2014). Do conservation measures in forests work? A comparison of three area-based conservation tools for wood-living species in boreal forests. *Forest Ecology and Management, 330*, 8–16.
40. Tambi, M. D., & Bime, W. J. (2019). Adequate financing and agricultural production: Response to new generation agriculture in Cameroon. *JASc (Journal of Agribusiness Sciences), 3*(1), 1–15.
41. Tracewski, Ł., Butchart, S. H., Donald, P. F., Evans, M., Fishpool, L. D., & Buchanan, G. M. (2016). Patterns of twenty-first century forest loss across a global network of important sites for biodiversity. *Remote Sensing in Ecology and Conservation, 2*(1), 37–44.
42. Vihervaara, P., Auvinen, A. P., Mononen, L., Törmä, M., Ahlroth, P., Anttila, S., & Virkkala, R. (2017). How essential biodiversity variables and remote sensing can help national biodiversity monitoring. *Global Ecology and Conservation, 10*, 43–59.
43. Watson, J. E., Darling, E. S., Venter, O., Maron, M., Walston, J., Possingham, H. P., & Brooks, T. M. (2016). Bolder science needed now for protected areas. *Conservation Biology, 30*(2), 243–248.
44. Wilson, O. (2023). Putting nature centre stage? The challenges of 'mainstreaming' biodiversity in the planning process. *Journal of Environmental Planning and Management, 66*(3), 549–571.
45. Zafra-Calvo, N., Pascual, U., Brockington, D., Coolsaet, B., Cortes-Vazquez, J. A., Gross-Camp, N., & Burgess, N. D. (2017). Towards an indicator system to assess equitable management in protected areas. *Biological Conservation, 211*, 134–141.

Chapter 4

THE IMPACT OF ENVIRONMENTAL POLICY ON RENEWABLE ENERGY INNOVATION

Contents

INTRODUCTION

Given the current global carbon crisis and the declining supply of traditional energy sources, it is clear that there is an urgent need for energy innovation, specifically in the field of renewable energy (RE). The United Nations unveiled a list of 17 Sustainable Development Goals (SDGs) to address the climate change catastrophe in 2015. Of these, two, SDGs 7 and 13, specifically focus on developing alternative energy sources, emphasizing renewable energy (RE). In addition, the Paris Agreement sets a goal of limiting the rise in global temperatures to 1.5°C above pre-industrial levels. This emphasizes the importance of green innovation, particularly in renewable energy (Appiah et al., 2023). Therefore, it is crucial to prioritize the transition to a low-carbon economy through renewable energy innovation. This is necessary to achieve the Sustainable Development Goals and meet the Paris Agreement objectives. Although the importance of renewable energy technology in combating climate change is clear, its implementation in real-world situations remains restricted the disparity highlighted here emphasizes the pressing need for innovation, specifically in addressing key challenges like an unstable electricity supply and expensive production costs. These factors persistently impede the adoption of renewable energy. Given the current circumstances, it is clear that there is an urgent need for more research on the elements that influence renewable energy innovation (Zhang et al., 2022).

It is widely acknowledged that market supply plays a crucial role in driving innovation. Nevertheless, depending exclusively on the market carries inherent market failure risks, which renders it insufficient to guarantee the ongoing advancement of innovation. Therefore, the inclusion of non-market factors becomes crucial in promoting and accelerating innovation, with a specific focus on the crucial role of government in implementing tangible policy interventions and stimulating investments (Nesta et al., 2014).

Environmental policies aim to shift resources towards cleaner alternatives and to avoid polluting energy sources. Nevertheless, the effectiveness of these policies in stimulating renewable energy innovation is still inconclusive and continues to be a topic of current discussion. Specifically, studies analyzing the impacts of environmental policies produced a diverse array of conflicting results (Usman et al., 2020).

There is a consensus among scholars that environmental policies substantially impact innovation in renewable energy. Environmental regulations can actively encourage the adoption of green innovations, particularly RE advances, according to Porter and van der Linde, (1995). It is also anticipated that the advantages of innovation will outweigh the expenses associated with compliance (Mounicou et al., 2009).

However, the increasing impact of climate change in recent years has raised significant concerns about the effectiveness of policy intervention in promoting the transition to renewable energy. Environmental policies can have a significant impact on the innovative behaviors of companies. It is worth mentioning that renewable energy technology often requires significant initial investments, which may discourage companies from adopting it. Furthermore, the burdensome compliance costs linked to rigorous environmental policies place additional financial pressure on companies, impeding their capacity to allocate resources toward innovative initiatives. In addition, companies that are subject to strict policies may choose to move their investments to other locations (Pitelis, 2018).

This work examines how environmental policies affect RE innovation in light of the contradictory results and how well they support it. Specifically, compiling the most recent research findings aims to highlight gaps and bring new insights. The main objective is to thoroughly analyze existing knowledge, pinpoint areas that need more research, and provide valuable insights into renewable energy's innovation and policy aspects (Widom, 1977).

4.1. RENEWABLE ENERGY INNOVATION

Given the increasing levels of atmospheric CO_2, largely due to the burning of fossil fuels and the effects of climate change, it is clear that we must prioritize the shift to cleaner energy sources. Amidst these challenges, there is a growing momentum behind the idea of a low-carbon economy. This approach emphasizes using environmentally friendly energy sources, known as RE, in consumption and production processes (Wüstenhagen et al., 2007).

According to Wang et al. (2019), renewable energy is derived from continuous and naturally repetitive energy flows in the local environment. This technology has been widely recognized for its low carbon footprint and has shown improved cost competitiveness lately. Companies demonstrate different strategic responses to shift towards renewable energy sources. Some take a proactive approach by engaging in renewable energy innovation. Innovation plays a crucial role in driving economic growth and achieving competitive advantages. There is a clear definition of innovation, specifically in renewable energy. It refers to the processes involved in inventing and improving new energy technologies for commercial purposes. Renewable energy innovation is highlighted by its ability to tackle the climate crisis by reducing emissions and pollution, conserving production resources, and fostering the development of clean energy technologies (Alvarez et al., 2017). RE innovation changes the production and distribution of energy, which substantially impacts the energy market supply. With the emergence of renewable energy technologies, business models and industry practices are undergoing significant changes to address environmental concerns and meet evolving energy needs. Climate change, including water and temperature availability shifts, requires using renewable energy technology as a sustainable option. This, in turn, impacts electricity supply systems and how they function (Bayer et al., 2013).

At the same time, renewable energy innovations transform people's energy consumption. With the rise of technological advancements like efficient home solar systems, consumer preferences are increasingly leaning toward sustainable energy solutions. This shift in demand is creating a growing market demand for renewable energy technology. In addition, various regional factors, such as societal attitudes and environmental regulations toward green energy, play a crucial role in shaping the adoption and diffusion of renewable energy innovations.

The various advancements in the supply and demand aspects highlight the complex nature of renewable energy innovation. It influences modifications in energy distribution strategy and production technology from the supply side. Regarding the demand side, it significantly influences energy use patterns and consumer preferences. Societal values and policy frameworks further shape these patterns. The combined effect of renewable energy innovation is vital in moving towards a more sustainable energy future, highlighting the interdependence of environmental policies, market dynamics, and technology (Yi et al., 2024).

However, renewable energy innovation does have inherent limitations. When comparing it to fossil fuels, some may find it less appealing because of the significant initial investments needed, the uncertainty that comes with the technology, the limited liquidity of assets, and the reliance on regulations. Given these factors, the shift to renewable energy may seem logical from an environmental standpoint. However, it could encounter financial and economic obstacles due to abundant affordable fossil fuels.

In addition, the trajectory of renewable energy (RE) innovation is shaped by two conflicting factors. On one hand, there is a critical urgency to address climate change by advancing clean energy generation. On the other hand, the widespread availability of inexpensive fossil fuel energy can hinder the growth of RE usage and production. Therefore, the progress of renewable energy innovation requires incentives beyond market forces, particularly through implementing environment-focused policies (Feng and Zheng, 2022).

Recognizing the inherent uncertainty in innovation and the acceptance of new technology, it becomes crucial to establish policies that effectively address and mitigate this uncertainty, thereby promoting a conducive environment for innovation. Although environmental policies and RE innovation are complementary approaches to establishing a low-carbon economy, it is crucial to understand that they can impact one another. This attempt focuses on government involvement's impact on promoting or hindering renewable energy innovation, specifically environmental policies (Liang and Fiorino, 2013).

4.2. ENVIRONMENTAL POLICIES

Authorities have set up numerous policies to make switching to a low-carbon economy easier. Government intervention in promoting renewable energy innovation can have significant benefits in three unique methods. First, regulations like quotas and demand subsidies may increase the market for renewable energy (RE), raising the expected return on investments made in RE and encouraging innovation. Furthermore, innovation in the field of renewable energy is often accompanied by a considerable amount of uncertainty. Effective policies addressing uncertainty challenges are crucial for fostering innovation in the renewable energy sector. One such policy is the provision of subsidies for research and development (R&D). Thirdly, the development of renewable energy technology can be seen as a direct result of two specific market failures: knowledge and environmental externalities, as supported by theoretical foundations. As part of the policy portfolio, it is important to consider specific policies that address those failures, like R&D subsidies and a tradable permit scheme (Kozluk and Zipperer, 2015).

There are different classifications of environmental policies. Environmental policies can be divided into nonmarket-based or command-and-control, and market-based. Nonmarket-based policies include emission controls, voluntary instruments, and R&D funding for renewable energy projects, and Market-based policies include feed-in tariffs, trading plans, and taxes.

Nicolli and Vona (2016) built upon the above classification in a separate study. This comprehensive classification includes four main environmental policy groups. These groups include market-based instruments like carbon taxes and feed-in tariffs; command-and-control policies like renewable portfolio standards and laws; voluntary instruments like RE targets; and hybrid instruments that combine elements of command-and-control and market-based policies like RE quotas with certificate trading (Toomey et al., 2007).

A study examined the impact of regulations on innovation in the wind power and solar industry. They found an overall 11 policy instruments that they then grouped into six clusters: (i) targets: these comprise strategies and targets related to renewable energy; (ii) R&D support: these comprise programs for research, development, and deployment; (iii) quotas: these comprise RE quotas without and with certificate trading; (iv) feed-in tariffs: these comprise public competitive bidding, net metering, and premium feed-in tariffs or fixed rate; (v) fiscal incentives: these comprise low-cost loans, capital subsidies, public spending, tax reduction, and tax credits; and (vi) carbon trading: these comprise greenhouse gas certificate trading systems (Al-Najjar and Anfimiadou, 2012).

Similarly, Renn (2006) has proposed a new classification system that includes seven distinct categories of environmental policies. These classifications encompass a range of factors, such as renewable energy certificates with quotas, obligations imposed on suppliers to source from renewables, voluntary programs by different stakeholders at the national level, tax measures, investment incentives, feed-in tariffs, and government R&D expenditures on renewable energy technology.

It is evident from the literature that there is a need for an all-encompassing classification that includes every kind of environmental policy. This chapter aims to consolidate existing research on the impact of environmental policy on renewable energy innovation. This approach centers on establishing an inclusive classification encompassing key environmental policies. Policy classifications have been removed from our refined sample to ensure that every policy is properly considered. Furthermore, this work thoroughly classified environmental policies according to the most common policy types, building on the similarities in different policy classifications. Expanding on the previous literature, the classification includes six categories: carbon pricing, emissions trading, fiscal incentives, quotas, feed-in tariffs, and renewable energy targets (Nesta et al., 2014).

Targets related to RE technology involve the development of specific goals and strategies. These publicly announced strategies demonstrate a strong dedication to expanding renewable energy and are widely regarded as favorable in this area. Feed-in tariff policies are specifically crafted to encourage renewable energy generation. These policies require the government to ensure that renewable energy producers receive prices higher than the market rate for a set period. Furthermore, this reduces the unpredictability of forthcoming electricity costs. Feed-in tariffs can encompass a range of tariff structures, including premium and fixed rates. They can also incorporate innovative mechanisms like tendering and net metering.

Quotas, referred to as Renewable Portfolio Standards (RPS), are a policy instrument requiring electricity retailers to obtain a specific or growing percentage of their electricity from renewable energy sources. Quotas, whether they involve certificate trading or not, can have significant impacts on R&D activities, which affects innovators in turn (Albrizio et al., 2017).

Fiscal incentives are another class of policy that comes from investment-based tools. Governments typically design these incentives to help reduce the upfront costs of adopting renewable energy technologies. Various incentives are available to support environmental initiatives, such as energy production payments, investment tax credits, subsidies, and government grants.

Emissions trading is another category of environmental policy. This category includes a variety of schemes, including price ceiling programs or quantity-based. Some noteworthy examples of these initiatives include the European Union's Emissions Trading System (EU ETS), the Cap-and-trade program in Quebec province, Canada, and the Carbon Trading Policy (CTP) in China (Bryman, 2008).

Carbon pricing, the category of environmental policies, raises the cost of fossil fuels to deter production. This strategy seeks to encourage the reduction of carbon emissions in the economy. Bubna-Litic and Stoianoff (2014) suggest that carbon pricing has three main effects on innovative low-carbon practices. First, it can place significant financial burdens on carbon-intensive companies, encouraging them to shift towards low-carbon technologies to evade the hefty carbon taxes. It is worth mentioning that for carbon pricing to encourage renewable energy innovation effectively, it needs to meet a specific threshold. Secondly, implementing carbon pricing can catalyze government incentives and support to stimulate private sector investment in innovative technologies. Finally, it has the potential to effectively limit the actions of companies that are not environmentally conscious, which could result in their downfall. This would create space for more innovative companies to step in and take over their market share (McCann et al., 2005).

4.3. INDIVIDUAL-POLICY LEVEL

This subsection thoroughly examines and explores the current research on the six environmental policy types developed at the individual policy level. These policies include carbon pricing, emissions trading, fiscal incentives, quotas, feed-in tariffs, and RE targets (Stadelmann and Eder, 2021).

4.3.1. RE Targets

Based on their research found that RE targets have had a significant influence on promoting innovation in the field of wind energy. This notable impact can be attributed to the declaration of renewable energy targets, which frequently do not align with the execution of particular environmental policies. On the other hand, a recent study by (Hille and Oelker, 2023) examined the impact of environmental policies on solar and wind energy. Surprisingly, their findings suggest that commonly used instruments like renewable energy targets may not be as effective in promoting innovation in renewable energy.

4.3.2. Feed-in Tariffs

There are varying results across different regions and studies when examining the impact of feed-in tariff policies. Frondel et al. (2010) contended that the feed-in tariff in Germany, a type of RE-supportive policy, is insufficient to drive technological innovation in the country. According to their study, German environmental policies do not provide sufficient incentives for renewable energy innovation. However, government policies and regulations were shown to have stimulated RE innovation since 1990 in a qualitative study carried out between 2009 and 2016 in the United Kingdom (Raybould et al., 2020). According to the study, the feed-in tariff policy from 2009 to 2016 resulted in a higher adoption of renewable energy. Premium tariffs and fixed rates were more effective in promoting renewable energy innovation than other methods like public competitive bidding and net metering. There may be a correlation between the timing of policy execution and the effectiveness of its implementation. Classic feed-in tariff instruments were established early in solar and wind technology development. At the same time, advanced versions were generally enacted during or after the height of technological advancement in these fields. According to a study by Hille et al. (2020), the worldwide effect of feed-in tariff policies shows significant variation, with the wind energy sector experiencing the most substantial effects. In a study conducted by Nicolli and Vona (2016), it was discovered that the feed-in tariff notably affected specifically solar energy technology in 19 countries. These varied findings emphasize the significance of considering technological and regional differences when evaluating the impact of feed-in tariff policies on promoting renewable energy innovation (Kim and Lee, 2012).

4.3.3. Quotas

The topic of the efficacy of quotas in promoting renewable energy innovation has been widely discussed in the literature. According to Dahlerup and Freidenvall (2010), mandates could offer stronger incentives for minor renewable energy innovation than carbon taxes. Nevertheless, if it involves radical innovation, carbon taxes are a more favorable choice as their effectiveness tends to diminish. The findings suggest that mandates may not effectively stimulate market-based innovation. In their study, (Hille et al., 2020) further contributed to this discussion by analyzing the solar and wind industries. It has been found that quotas specific to technology have a limited impact on promoting innovation. It was noted that tradable renewable certificates primarily benefit price-competitive and mature technologies like biomass, geothermal, and hydro, resulting in a limited overall impact.

In their study, Nicolli and Vona (2016 examined the diverse effects of renewable energy policies on innovation, highlighting that these effects differ depending on the level of development of each technology. Wind and other mature forms of renewable energy often see advantages from implementing quotas, as they tend to have lower compliance costs. On the other hand, Hille and Oelker (2023) put forth the viewpoint that quotas with certificate trading have the potential to be highly effective in promoting low-carbon innovation.

A study by Krook et al. (2009) focused on analyzing how carbon reduction mandates affect firms' deployment of low-carbon innovation. Findings indicate that carbon reduction mandates can stimulate low-carbon innovation, although the extent of this innovation may vary. State-owned firms positively impact the low-carbon patents ratio, whereas non-state-owned firms demonstrate an effect on the number of patents. In addition, the impact of low-carbon innovation differs depending on the type, with low-quality innovation having a more pronounced effect than deep innovation (Murray, 2014).

These findings highlight the importance of considering different factors, such as innovation type, technology maturity, and policy design, when evaluating the impact of quotas on promoting low-carbon innovation (Krook and Norris, 2014).

4.3.4. Fiscal Incentives

The literature highlights the important role of fiscal incentives in promoting renewable energy innovation, especially considering the large initial investments needed for renewable energy technologies. Government subsidies can potentially encourage new players to enter the market, intensifying competition. This competition has the potential to inspire firms to embrace risks and adopt innovative practices. Countries with liberalized energy markets show a more pronounced positive impact of RE-supportive fiscal policies, as shown by (Hall and Van Reenen, 2000).

According to Nicolli and Vona (2016), fiscal incentives such as government R&D expenditure and demand subsidies have a greater impact on promoting innovation in less mature renewable energy technologies. Nevertheless, Fox and Murray (2004) highlighted the noteworthy influence of fiscal policies that support renewable energy on various technologies, including well-established ones like wind energy. A recent study discovered that fiscal incentives have proven to be highly effective in promoting growth within the solar and wind energy sectors.

Focusing on governmental R&D expenditure, a study analyzed the effects on innovation in renewable energy and fossil fuel sectors using data from Chinese provinces. It has been shown that government R&D expenditure, specifically in environmental regulations, can effectively drive innovation. The impact of this phenomenon is substantial for renewable energy sources but minimal for fossil fuel technologies. In addition, a recent study in China has provided further evidence supporting the Porter hypothesis. The study demonstrates that climate policies, such as government R&D investments, drive low-carbon innovation.

In a separate study, Lim (1983) examined the impact of public research and development spending on advancing renewable energy technologies between 2000 and 2015. Using country-level data, their research has shown that these policies effectively drive renewable energy innovation. The policy's effectiveness, however, differs among various renewable energy sources (wind, solar, and biomass energy) and countries (Jin et al., 2005).

4.3.5. Emissions Trading

Extensive research has been conducted on the impact of emissions trading policies, including the carbon trading schemes and EU ETS, on low-carbon and renewable energy innovation. In their study, employed the difference-in-difference approach to examine the impact of the EU ETS, a well-known cap-and-trade system, on the rate of low-carbon patenting. It was discovered that the EU ETS substantially impacted the promotion of low-carbon innovation within firms regulated by the EU ETS. This resulted in a noteworthy 10% increase in their overall innovation output. Compared to similar non-EU ETS firms, the impact of the EU ETS was relatively weaker, leading to a mere 1% increase in innovation. According to a study, the EU ETS had a minor effect on promoting renewable energy innovation.

However, several other studies have shown the positive impact of emission trading policies on promoting innovation. According to the EU ETS has been found to promote low-carbon innovation, such as renewable energy, by boosting the number of low-carbon patents and increasing investment in research and development. Muller and Mestelman (1998) examined the pilot carbon trading policy of China and emphasized the significant impact of environmental policies on driving renewable energy innovation. Research has demonstrated that emission trading schemes positively impact low-carbon innovation in different sectors. This effect strengthens over time as financial constraints are alleviated. A study emphasized the importance of market characteristics in determining the strength of the relationship. They also highlighted the positive effect of stringent policies in driving low-carbon innovation. Zhang et al. (2014) examined the EU ETS. They investigated if there was a substantial impact on the effectiveness of the various temporal stages of the EU ETS. Using the Schumpeterian "creative response" framework, the authors evaluated the firm's approach to green policies and their impact on green innovation. Using innovative methods to conduct business is part of this creative response. Based on their research, it was observed that firms operating in energy-intensive industries showed a creative response to the EU ETS, specifically in terms of environmental innovation. This implies that businesses in these industries used innovative strategies and practices in response to the EU ETS, fostering ecologically conscious and green innovation.

The study conducted by Baldwin (2008) aimed to explore the impact of China's national carbon emission scheme on the advancement of low-carbon technology. In contrast to numerous other studies in this domain, Liu et al. (2015) adopted a unique methodology instead of relying on patenting as a metric for renewable energy innovation. They analyzed the correlation between low-carbon innovation output and input while considering environmental constraints to assess innovation efficiency. According to their findings, there was a three-year lag in the influence of the carbon emission system on innovation. On the other hand, their findings diverged from previous studies as they did not support a swift advancement in renewable energy innovation. The impact of environmental policies, specifically carbon emission schemes, on the development of low-carbon technology may differ in intensity and timing (Bayer and Aklin, 2020).

4.3.6. Carbon Pricing

Carbon pricing policies, such as carbon taxes and additional measures, may impact renewable energy innovation differently. Research in this area has yielded conflicting results. The success of these policies relies on a range of factors, such as the particular circumstances and the prospects for research and development.

In a study conducted by Klenert et al. (2018), the researchers examined how institutional determinants influenced renewable energy innovation in various countries. It was discovered that higher oil prices had the potential to drive innovation in renewable energy, although the impact varied from country to country. The impact was less pronounced in OECD countries than non-OECD countries, with certain renewable energy technologies, like solar and wind innovation, experiencing a more substantial effect.

On the other hand, a study by Tvinnereim and Mehling (2018) analyzed how energy prices can serve as a policy tool to stimulate innovation in renewable and fossil fuel sectors. It was discovered that fluctuations in energy prices had a greater effect on fossil fuel technologies than renewable energy technologies.

This suggests that the influence of carbon pricing can differ depending on the technology and energy source.

According to a study by Stavins (2020), carbon taxes were one of the worst-performing policy instruments in the solar and wind energy sectors. The impact of these taxes on promoting renewable energy innovation throughout these sectors was minimal. Best et al. (2020) conducted a study comparing the effects of carbon taxes and mandates on innovation. They discovered that the influence of a carbon tax on renewable energy innovation varied depending on the prospects for research and development. In situations where the prospects for R&D were promising, carbon taxes proved to be more effective in promoting renewable energy innovation than mandates. However, this dynamic was reversed when the outlook for R&D was less optimistic.

Baranzini et al. (2017) conducted a comprehensive study on the effects of various environmental policies, such as regulations and carbon taxes, on renewable energy innovation in the electricity industry across 27 OECD countries. It was discovered that carbon taxes proved to be more successful in encouraging investment in renewable energy technology compared to regulatory standards. Nevertheless, the impact of carbon taxes on renewable energy innovation varied across different countries, indicating that national contexts influenced their effectiveness.

Several studies in the sample analyzed the impact of general environmental policies on fostering renewable energy innovation. These studies did not prioritize the policy types mentioned earlier, so their findings are presented separately. According to Aldy and Stavins (2012), these policies positively impact RE innovation, highlighting the significance of frameworks supporting policy in promoting innovation.

On the other hand, Trinks et al. (2022) offered an alternative viewpoint. Through evaluating renewable energy policies across countries, it was discovered that there is a lack of emphasis on technology-related indicators, which are essential for driving renewable energy innovation. There appears to be a possible adverse correlation between the development of renewable energy technology development and policy measures. According to Campiglio (2016), policymakers should prioritize adopting and developing renewable energy innovation.

In a recent study, Best et al. (2020) delved into the impact of environmental policies on domestic economies, exploring both the indirect and direct effects on innovation. The findings were varied, as renewable energy policies were found to dampen innovation in most of the countries studied, except for Germany, Japan, the USA, Japan, and Germany. According to Baldwin (2008), there is a theory called the "California effect," which suggests that the foreign environmental policies enacted by trading partner countries can significantly impact domestic firms' innovation in renewable energy through international trade. The influence of these policies could, at times, exceed the impact of domestic measures. These studies highlight the complex and situation-specific connection between renewable energy innovation and general environmental policies. Additional research is necessary to develop a more holistic understanding of this relationship.

4.4. POLICY-MIX LEVEL

The previous subsection examined the literature regarding the direct impacts of various environmental policies. This subsection thoroughly examines and combines the different elements of diverse environmental policies. By considering how these policies interact, you can comprehensively understand their combined impact on renewable energy innovation. We also evaluate the variations between these findings and those associated with each policy.

There are two primary reasons why Bouma et al. (2019) suggest examining a policy mix instead of focusing on individual programs. Firstly, technological change encompasses various risks and market failures, necessitating a comprehensive policy approach beyond a single solution. Numerous scholars have also highlighted the significance of policy mixes in tackling market failures like information asymmetry and externalities. Secondly, it is important to note that policies are not present in isolation; they evolve with existing policies. Therefore, it is important to consider a combination of policies rather than relying on a single policy to fully understand the impact on institutions. It is crucial to consider the timing of policies, as it can significantly impact the direction and magnitude of the effects on renewable energy innovation (Flanagan et al., 2011).

Combining different policies is a common approach to tackling intricate problems. When creating these combinations, carefully considering both the conflicting and complementary impacts is important. In a policy mix, the effectiveness of every individual policy is not the only factor to consider. It is also important to examine how these policies interact, whether they conflict, work together synergistically, or complement each other. These interactions can significantly impact the policies' general efficacy as a whole. Therefore, a crucial question comes up: which policy mix demonstrates the highest level of effectiveness? Several policy combinations researchers have looked into will be explored using the extant body of literature. It is important to mention that the studies presented in this subsection primarily concentrate on policy mixes at a cluster level instead of individual policy types (Bahn and Wilts, 2017). Technology-push and market-pull policies must work together, as Vītola (2015) emphasized. Without market-pull policies that stimulate demand for renewable energy technologies, technology-push policies that encourage technological innovation have limited effectiveness. For instance, supporting policies for renewable energy could impact consumer behavior without affecting supply behavior. When demand for fossil fuels decreases, it can have the unintended consequence of lowering prices. This, in turn, stimulates the supply of fossil fuels and increases their demand, which goes against the goals of environmental policies.

In recent studies, researchers have discovered that combining policies supporting renewable energy can greatly enhance their positive effects on innovation in this field. These findings are supported by multiple studies conducted by Shao and Huang, (2023), which suggest that a diverse set of instruments that support renewable energy is the most effective approach when designing a policy mix. Nevertheless, the effectiveness of market-based policies in driving innovation was challenged researchers who made a case that the lack of comprehensive energy pricing mechanisms limits the effectiveness of market-pull policies in driving innovation, as they primarily impose cost pressures on firms. However, it has been confirmed that technology-push policies have a significant innovation effect, particularly in high-income and OECD countries.

Ali et al. (2023) executed a thorough analysis of different policy combinations in the wind energy sector. These policies were classified into three main groups. Market-based policies are represented by Demand-side policies (DSP) and Supply-side policies (SSP), while Environmental-side policies (ESP) encompass non-market-based policies. The study's findings validate the positive interactions observed between different combinations of mixes. Essentially, the interplay between different policies can help resolve uncertainties and conflicts, leading to a notable boost in renewable energy innovation. These findings are consistent with the research conducted by (Liu et al., 2022). On the other hand, Wang et al. (2019) discovered a contradictory impact between policies, specifically between ESP and DSP, which needs to be addressed and resolved.

Shao and Huang (2023) further supported that combining various policy types can promote renewable energy innovation. Nevertheless, the success of a policy mix in promoting innovation relies heavily on the specific features and design of each policy. Based on survey data from the German renewable energy sector, Wang et al. (2019) conducted a study to examine the impact of policy design features on renewable energy innovation and the correlation between a policy mix and the stimulation of innovation. They examined four important aspects of a policy mix: coherence, comprehensiveness, credibility, and consistency. The study examines the positive effects of a policy mix on low-carbon innovation and emphasizes the importance of credibility and consistency within such a mix.

Bubna-Litic and Stoianoff (2014) investigated how greenhouse gas pricing interacts with policies that encourage RE innovation in the UK, Canada, and Australia in a different study. These findings suggest that the pricing of greenhouse gases can incentivize the development of creative strategies to address climate change. It is important to consider that without carbon pricing, relying solely on supportive policies may not be sufficient to drive advancements in renewable energy technology.

4.5. DISCUSSION, FUTURE RESEARCH, AND IMPLICATIONS

The previous section thoroughly examined the existing literature to understand the impact of different environmental policies on renewable energy innovation. The following discussion will delve into the evolution of research on the relationship between renewable energy innovation and environmental policies since 2005. In addition, it provides a comprehensive analysis of the practical and theoretical implications of the review, offering valuable insights for policymakers and contributing to the wider academic discussion on renewable energy innovation and environmental policy (Lee and Choi, 2011).

4.5.1. Discussion

It is widely recognized that policymakers have a crucial role in driving the shift toward a low-carbon economy and advancing the adoption of clean energy technology. These policymakers have a significant impact by creating effective environmental policies. It is worth mentioning that the impact of environmental policies can differ depending on their specific types (Parker and Hess, 2001).

This analysis utilized a two-tier approach, carefully examining policies at the individual level and policy combinations to evaluate their influence on renewable energy innovation. Examining individual policies allows for a comprehensive grasp of their specific impacts, as highlighted by recent research. However, it is important also to consider the potential synergistic effects that can emerge when multiple policies are implemented together, as demonstrated by previous studies. As emphasized by Blum (2002) policy-mix analysis effectively captures the synergies between different policies and helps minimize uncertainties that may arise from implementing individual policies. Furthermore, it considers the intricate relationships between various policies.

Policymakers need to consider the regulatory context and objectives carefully when making policy decisions. They should know the trade-offs and benefits of different approaches.

This analysis thoroughly examined the literature on six different policy types at the individual-policy level. Although there is agreement in the literature about the beneficial effects of fiscal incentives on renewable energy innovation, the research on other policy types has produced uncertain results.

Previous research on the impact of renewable energy targets on renewable energy innovation is lacking, and the findings from the Bass (1949) study seem inconclusive. In addition, Hille and Oelker, 2023) have put forth the argument that the maturity of policies has a negative impact on their effectiveness. On the other hand, according to Hille et al. (2020), older, more established feed-in tariff models have proven more effective than newer, advanced versions. Academics are at odds on how feed-in tariffs affect the innovation of renewable energy sources. This matter lacks agreement due to differences among renewable energy technologies and countries.

There is a lack of consensus in the literature regarding the effectiveness of quotas. Several studies have examined the effects of quotas on renewable energy (RE) innovation. Hille and Oelker (2023) have presented evidence supporting the positive impacts of quotas on RE innovation. On the other hand, Bloom (1953) has found that the impact of quotas on RE innovation is not significant. According to the research that is currently available, emission trading systems have a positive impact on RE innovation. However, the research conducted by Chinn et al. (2001) found that this positive effect is limited to regulated companies and may not apply to all businesses. In addition, the literature review on carbon pricing reveals various findings regarding the impact of carbon taxes and energy prices on innovation. Various studies have presented mixed results.

In terms of policy composition, researchers have shown that combining environmental policies in a way that creates synergy can amplify the innovation impact of the institutional framework According to Lee and Choi (2011), combining technology-push and market-pull policies is important to maximize their innovation effect. This combination supports the supply and demand sides of renewable energy innovation. On the other hand, Zhang et al. (2022) argued that market-pull policies might not yield the desired results in this situation.

4.5.2. Future Research Agenda

Based on this discussion, it is evident that further research should focus on various important avenues to address the current literature gaps. This includes a detailed analysis of the effects of Renewable energy targets that go beyond the solar and wind sectors that are regularly studied. Various studies, including those by (Welbourne and Mejia, 1995), have yielded inconclusive findings, highlighting the importance of further investigation into a wide range of renewable energy sources. Based on research conducted in different countries, it is clear that feed-in tariffs require further investigation to determine their effectiveness. It would be beneficial for future research to examine the specific effects of these policies on various renewable energy technologies in different countries. The importance of quotas in renewable energy innovation should not be overlooked. There are differing perspectives in the literature regarding the effectiveness of newer technologies. Certain studies suggest limited effectiveness, while others provide contradictory evidence. Further research is necessary to reach a more conclusive understanding (Paul et al., 2023).

In addition, a more thorough examination is needed for emission-trading schemes, particularly the EU ETS and China's CTP. Although the impact of the CTP on RE innovation has been well-documented by various studies, the EU ETS has yielded mixed results, as highlighted by research conducted by (Marlow, 2014), which suggests the need for further investigation. Furthermore, there is a lack of research on the direct impact of carbon taxes on renewable energy innovation. There is a noticeable gap in research regarding the standalone effects, which calls for more focused studies.

In addition, the current literature does not sufficiently address the interactions inside policy mixes and their general effect on renewable energy innovation. Further research should analyze those combinations to identify successful synergies. Finally, the environmental policy mixes design and their impact on innovation yield inconclusive results. Different studies have conflicting findings regarding the impact of various policy types on a mix). There seems to be a lack of consistency in the approach and a restricted application of qualitative methods. This suggests that there is room for more extensive research, possibly employing various methodologies to obtain a more comprehensive understanding (Bloom, 1953).

4.5.3. Theoretical Implications

This work adds to the growing knowledge of renewable energy innovation and environmental policy from a theoretical standpoint. The multifaceted and complex nature of policy impacts on renewable energy innovation is highlighted, which questions the conventional perception of policy effectiveness in isolation. Our analysis supports a comprehensive understanding of the effects of policies, considering the interaction between various policy tools and their combined impact on renewable energy innovation (Sexton and Whiston, 1994). More advanced theoretical models are required to accurately depict the complexities of policy mixes and how they interact. In addition, this attempt points out deficiencies in existing theoretical frameworks concerning the precise effects of various policy types, such as quotas, feed-in tariffs, and renewable energy targets on renewable energy innovation. Further theoretical development is necessary to account for the diverse impacts of these policies in different national and technological settings (Sarason et al., 1987).

4.5.4. Practical Implications

The findings from this review have significant practical implications for industry stakeholders and policymakers in the field of renewable energy innovation. Having a deep understanding of the intricate impacts of different environmental policies is essential when it comes to crafting robust policy frameworks. For example, current analysis indicates that the impact of feed-in tariffs and RE targets varies greatly depending on the national context and the specific renewable energy technologies being considered. Policymakers need to customize these policies to meet their renewable energy sectors' unique requirements and development stages. It is clear that the effectiveness of emission trading schemes and quotas, like China's CTP and EU ETS, varies. This highlights the need for ongoing policy adaptation and evaluation. This emphasizes the importance of adopting a flexible policy approach that can adapt to evolving technology and market conditions (Phillips, 1984).

Furthermore, the lack of extensive research on the direct impact of carbon taxes on renewable energy innovation highlights the importance of further policy investigation in this area. Policymakers need to conduct focused studies to understand better how carbon taxes can promote innovation in renewable energy. The findings also indicate the need for a balanced combination of policies, incorporating different types of policies to foster an environment that promotes renewable energy innovation. Policymakers must prioritize the development of comprehensive policy portfolios that effectively tackle the unique opportunities and challenges within renewable energy sectors. This approach will encourage continuous growth and innovation in renewable technologies (Cepeda et al., 2009).

4.6. SUMMARY

There has been a noticeable increase in study interest in RE innovation, focusing on the impact of environmental policies on impeding or motivating innovation. This is a comprehensive effort to bring together and analyze the current research in this field. The main goal here was to pinpoint the deficiencies in this study area, particularly emphasizing different types of policies. Using an SLR methodology, a comprehensive set of six policy types has been developed to cover all aspects of environmental policies. The proposed classification is extremely important, considering the current literature categorizations are still inconclusive. Through a comprehensive synthesis and analysis of the categorizations in the existing literature, this classification has been developed. This classification encourages us to differentiate between different policies according to how they affect RE innovation.

In addition, the conclusions of the current literature from individual policy and the policy-mix level have been thoroughly analyzed. This analysis has uncovered various environmental policies' impacts on renewable energy innovation. Nevertheless, there seems to be a prevailing agreement in the literature regarding specific policy types like fiscal incentives.

There are certain limitations to consider in this literature review. To ensure the highest quality of our final sample, we have chosen only to include academic journal articles and exclude sources such as reports, books, and conference papers. Although this approach improves the thoroughness of the review, it may have led to the exclusion of possibly important information from these other sources. In addition, the studies were chosen based on specific databases and keywords. Therefore, it is important to note that additional studies may not be included in our analysis. These studies may have used different terminology or been published in less well-known academic journals.

REFERENCES

1. Albrizio, S., Kozluk, T., & Zipperer, V. (2017). Environmental policies and productivity growth: Evidence across industries and firms. *Journal of Environmental Economics and Management, 81*, 209–226.
2. Aldy, J. E., & Stavins, R. N. (2012). The promise and problems of pricing carbon: Theory and experience. *The Journal of Environment & Development, 21*(2), 152–180.
3. Ali, A., Hasan, Z. U., Abbasi, Q., & Sulehri, F. A. (2023). Business or politics: Exploring the determinants of policy mix in South Asia. *Bulletin of Business and Economics (BBE), 12*(3), 114–123.
4. Al Najjar, B., & Anfimiadou, A. (2012). Environmental policies and firm value. *Business Strategy and the Environment, 21*(1), 49–59.
5. Alvarez-Herranz, A., Balsalobre-Lorente, D., Shahbaz, M., & Cantos, J. M. (2017). Energy innovation and renewable energy consumption in the correction of air pollution levels. *Energy Policy, 105*, 386–397.
6. Appiah, M., Li, M., Naeem, M. A., & Karim, S. (2023). Greening the globe: Uncovering the impact of environmental policy, renewable energy, and innovation on ecological footprint. *Technological Forecasting and Social Change, 192*, 122561.
7. Bahn-Walkowiak, B., & Wilts, H. (2017). The institutional dimension of resource efficiency in a multi-level governance system: Implications for policy mix design. *Energy Research & Social Science, 33*, 163–172.
8. Baldwin, R. (2008). Regulation lite: The rise of emissions trading. *Law and Financial Markets Review, 2*(3), 262–278.
9. Baranzini, A., Van den Bergh, J. C., Carattini, S., Howarth, R. B., Padilla, E., & Roca, J. (2017). Carbon pricing in climate policy: Seven reasons, complementary instruments, and political economy considerations. *Wiley Interdisciplinary Reviews: Climate Change, 8*(4), e462.
10. Bass, B. M. (1949). An analysis of the leaderless group discussion. *Journal of Applied Psychology, 33*(6), 527.
11. Bayer, P., & Aklin, M. (2020). The European Union emissions trading system reduced CO2 emissions despite low prices. *Proceedings of the National Academy of Sciences, 117*(16), 8804–8812.
12. Bayer, P., Dolan, L., & Urpelainen, J. (2013). Global patterns of renewable energy innovation, 1990–2009. *Energy for Sustainable Development, 17*(3), 288–295.
13. Best, R., Burke, P. J., & Jotzo, F. (2020). Carbon pricing efficacy: Cross-country evidence. *Environmental and Resource Economics, 77*(1), 69–94.

14. Bloom, B. S. (1953). Thought-processes in lectures and discussions. *The Journal of General Education, 7*(3), 160–169.
15. Blum, W. (2002). ICMI Study 14: Applications and modelling in mathematics education–Discussion document. *Educational Studies in Mathematics, 51*(1), 149–171.
16. Boell, S. K., & Cecez-Kecmanovic, D. (2014). A hermeneutic approach for conducting literature reviews and literature searches. *Communications of the Association for Information Systems, 34*(1), 12.
17. Bouma, J. A., Verbraak, M., Dietz, F., & Brouwer, R. (2019). Policy mix: Mess or merit? *Journal of Environmental Economics and Policy, 8*(1), 32–47.
18. Bryman, A. (2008). Of methods and methodology. *Qualitative Research in Organizations and Management: An International Journal, 3*(2), 159–168.
19. Bubna-Litic, K., & Stoianoff, N. P. (2014). Carbon pricing and renewable energy innovation: A comparison of Australian, British, and Canadian carbon pricing policies. *Environmental and Planning Law Journal, 31*(5), 368–384.
20. Campiglio, E. (2016). Beyond carbon pricing: The role of banking and monetary policy in financing the transition to a low-carbon economy. *Ecological Economics, 121*, 220–230.
21. Cepeda, N. J., Coburn, N., Rohrer, D., Wixted, J. T., Mozer, M. C., & Pashler, H. (2009). Optimizing distributed practice: Theoretical analysis and practical implications. *Experimental Psychology, 56*(4), 236–246.
22. Chinn, C. A., Anderson, R. C., & Waggoner, M. A. (2001). Patterns of discourse in two kinds of literature discussion. *Reading Research Quarterly, 36*(4), 378–411.
23. Clayton, A. (2021). How do electoral gender quotas affect policy? *Annual Review of Political Science, 24*(1), 235–252.
24. Dahlerup, D., & Freidenvall, L. (2010). Judging gender quotas: Predictions and results. *Policy & Politics, 38*(3), 407–425.
25. Devillé, W. L., Buntinx, F., Bouter, L. M., Montori, V. M., De Vet, H. C., Van der Windt, D. A., & Bezemer, P. D. (2002). Conducting systematic reviews of diagnostic studies: Didactic guidelines. *BMC Medical Research Methodology, 2*, 1–13.
26. Feng, G. F., & Zheng, M. (2022). Economic policy uncertainty and renewable energy innovation: International evidence. *Innovation and Green Development, 1*(2), 100010.

27. Flanagan, K., Uyarra, E., & Laranja, M. (2011). The 'policy mix' for innovation: Rethinking innovation policy in a multi-level, multi-actor context. *Research Policy, 40*(5), 702–713.

28. Fox, W. F., & Murray, M. N. (2004). Do economic effects justify the use of fiscal incentives? *Southern Economic Journal, 71*(1), 78–92.

29. Frondel, M., Ritter, N., Schmidt, C. M., & Vance, C. (2010). Economic impacts from the promotion of renewable energy technologies: The German experience. *Energy Policy, 38*(8), 4048–4056.

30. Hall, B., & Van Reenen, J. (2000). How effective are fiscal incentives for R&D? A review of the evidence. *Research Policy, 29*(4, 5), 449–469.

31. Hassan, M., Mattingly, D., & Nugent, E. R. (2022). Political control. *Annual Review of Political Science, 25*(1), 155–174.

32. He, X., Hong, W., Pan, X., Lu, G., & Wei, X. (2021). SARS-CoV-2 Omicron variant: Characteristics and prevention. *MedComm, 2*(4), 838–845.

33. Hille, E., & Oelker, T. J. (2023). International expansion of renewable energy capacities: The role of innovation and choice of policy instruments. *Ecological Economics, 204*, 107658.

34. Jardot, D., Eichhammer, W., & Fleiter, T. (2010). Effects of economies of scale and experience on the costs of energy-efficient technologies: Case study of electric motors in Germany. *Energy Efficiency, 3*, 331–346.

35. Jin, H., Qian, Y., & Weingast, B. R. (2005). Regional decentralization and fiscal incentives: Federalism, Chinese style. *Journal of Public Economics, 89*(9, 10), 1719–1742.

36. Kelly-Richards, S., Silber-Coats, N., Crootof, A., Tecklin, D., & Bauer, C. (2017). Governing the transition to renewable energy: A review of impacts and policy issues in the small hydropower boom. *Energy Policy, 101*, 251–264.

37. Kim, K. K., & Lee, C. G. (2012). Evaluation and optimization of feed-in tariffs. *Energy Policy, 49*, 192–203.

38. Klenert, D., Mattauch, L., Combet, E., Edenhofer, O., Hepburn, C., Rafaty, R., & Stern, N. (2018). Making carbon pricing work for citizens. *Nature Climate Change, 8*(8), 669–677.

39. Knox, K. S., Adams, J. R., Djulbegovic, B., Stinson, T. J., Tomori, C., & Bennett, C. L. (2000). Reporting and dissemination of industry versus non-profit sponsored economic analyses of six novel drugs used in oncology. *Annals of Oncology, 11*(12), 1591–1595.

40. Kozluk, T., & Zipperer, V. (2015). Environmental policies and productivity growth: A critical review of empirical findings. *OECD Journal: Economic Studies, 2014*(1), 155–185.

41. Krook, M. L., & Norris, P. (2014). Beyond quotas: Strategies to promote gender equality in elected office. *Political Studies, 62*(1), 2–20.

42. Krook, M. L., Lovenduski, J., & Squires, J. (2009). Gender quotas and models of political citizenship. *British Journal of Political Science, 39*(4), 781–803.

43. Lather, P. (2013). Methodology-21: What do we do in the afterward? *International Journal of Qualitative Studies in Education, 26*(6), 634–645.

44. Lee, Y., & Choi, J. (2011). A review of online course dropout research: Implications for practice and future research. *Educational Technology Research and Development, 59*, 593–618.

45. Liang, J., & Fiorino, D. J. (2013). The implications of policy stability for renewable energy innovation in the United States, 1974–2009. *Policy Studies Journal, 41*(1), 97–118.

46. Lim, D. (1983). Fiscal incentives and direct foreign investment in less developed countries. *The Journal of Development Studies, 19*(2), 207–212.

47. Liu, L., Chen, C., Zhao, Y., & Zhao, E. (2015). China's carbon-emissions trading: Overview, challenges and future. *Renewable and Sustainable Energy Reviews, 49*, 254–266.

48. Liu, Z., Wu, Z., & Zhu, M. (2022). Research on the green effect of environmental policies: From the perspective of policy mix. *Sustainability, 14*(23), 15959.

49. Marlow, S. (2014). Exploring future research agendas in the field of gender and entrepreneurship. *International Journal of Gender and Entrepreneurship, 6*(2), 102–120.

50. McCann, L., Colby, B., Easter, K. W., Kasterine, A., & Kuperan, K. V. (2005). Transaction cost measurement for evaluating environmental policies. *Ecological Economics, 52*(4), 527–542.

51. Mignon, I., & Kanda, W. (2018). A typology of intermediary organizations and their impact on sustainability transition policies. *Environmental Innovation and Societal Transitions, 29*, 100–113.

52. Mounicou, S., Szpunar, J., & Lobinski, R. (2009). Metallomics: The concept and methodology. *Chemical Society Reviews, 38*(4), 1119–1138.

53. Muller, R. A., & Mestelman, S. (1998). What have we learned from

emissions trading experiments? *Managerial and Decision Economics, 19*(4, 5), 225–238.

54. Murray, R. (2014). Quotas for men: Reframing gender quotas as a means of improving representation for all. *American Political Science Review, 108*(3), 520–532.
55. Nesta, L., Vona, F., & Nicolli, F. (2014). Environmental policies, competition and innovation in renewable energy. *Journal of Environmental Economics and Management, 67*(3), 396–411.
56. Nicolli, F., & Vona, F. (2016). Heterogeneous policies, heterogeneous technologies: The case of renewable energy. *Energy Economics, 56*, 190–204.
57. Parker, W. C., & Hess, D. (2001). Teaching with and for discussion. *Teaching and Teacher Education, 17*(3), 273–289.
58. Paul, J., Ueno, A., & Dennis, C. (2023). ChatGPT and consumers: Benefits, pitfalls and future research agenda. *International Journal of Consumer Studies, 47*(4), 1213–1225.
59. Phillips, J. K. (1984). Practical implications of recent research in reading. *Foreign Language Annals, 17*(4), 285–296.
60. Pitelis, A. T. (2018). Industrial policy for renewable energy: The innovation impact of European policy instruments and their interactions. *Competition & Change, 22*(3), 227–254.
61. Porter, M. E., & Linde, C. V. D. (1995). Toward a new conception of the environment-competitiveness relationship. *Journal of Economic Perspectives, 9*(4), 97–118.
62. Raybould, M. I., Kovaltsuk, A., Marks, C., & Deane, C. M. (2021). CoV-AbDab: The coronavirus antibody database. *Bioinformatics, 37*(5), 734–735.
63. Renn, O. (2006). Participatory processes for designing environmental policies. *Land Use Policy, 23*(1), 34–43.
64. Rowley, J., & Slack, F. (2004). Conducting a literature review. *Management Research News, 27*(6), 31–39.
65. Sandelowski, M., & Barroso, J. (2002). Finding the findings in qualitative studies. *Journal of Nursing Scholarship, 34*(3), 213–219.
66. Sarason, I. G., Sarason, B. R., Shearin, E. N., & Pierce, G. R. (1987). A brief measure of social support: Practical and theoretical implications. *Journal of Social and Personal Relationships, 4*(4), 497–510.
67. Sexton, T. L., & Whiston, S. C. (1994). The status of the counseling

relationship: An empirical review, theoretical implications, and research directions. *The Counseling Psychologist, 22*(1), 6–78.

68. Shao, J., & Huang, P. (2023). The policy mix of green finance in China: An evolutionary and multilevel perspective. *Climate Policy, 23*(6), 689–703.
69. Stadelmann-Steffen, I., & Eder, C. (2021). Public opinion in policy contexts: A comparative analysis of domestic energy policies and individual policy preferences in Europe. *International Political Science Review, 42*(1), 78–94.
70. Stavins, R. N. (2020). The future of US carbon-pricing policy. *Environmental and Energy Policy and the Economy, 1*(1), 8–64.
71. Stein, R. E., & Riessman, C. K. (1980). The development of an impact-on-family scale: Preliminary findings. *Medical Care, 18*(4), 465–472.
72. Talen, E. (1996). Do plans get implemented? A review of evaluation in planning. *Journal of Planning Literature, 10*(3), 248–259.
73. Toomey, T. L., Lenk, K. M., & Wagenaar, A. C. (2007). Environmental policies to reduce college drinking: An update of research findings. *Journal of Studies on Alcohol and Drugs, 68*(2), 208–219.
74. Tranfield, D., Denyer, D., & Smart, P. (2003). Towards a methodology for developing evidence-informed management knowledge by means of systematic review. *British Journal of Management, 14*(3), 207–222.
75. Trinks, A., Mulder, M., & Scholtens, B. (2022). External carbon costs and internal carbon pricing. *Renewable and Sustainable Energy Reviews, 168*, 112780.
76. Tvinnereim, E., & Mehling, M. (2018). Carbon pricing and deep decarbonisation. *Energy Policy, 121*, 185–189.
77. Usman, O., Alola, A. A., & Sarkodie, S. A. (2020). Assessment of the role of renewable energy consumption and trade policy on environmental degradation using innovation accounting: Evidence from the US. *Renewable Energy, 150*, 266–277.
78. Vītola, A. (2015). Innovation policy mix in a multi-level context: The case of the Baltic Sea Region countries. *Science and Public Policy, 42*(3), 401–414.
79. Wang, M., Wang, G., Sun, Z., Zhang, Y., & Xu, D. (2019). Review of renewable energy-based hydrogen production processes for sustainable energy innovation. *Global Energy Interconnection, 2*(5), 436–443.
80. Welbourne, T. M., & Mejia, L. R. G. (1995). Gainsharing: A critical review and a future research agenda. *Journal of Management, 21*(3), 559–609.

81. Widom, C. S. (1977). A methodology for studying noninstitutionalized psychopaths. *Journal of Consulting and Clinical Psychology, 45*(4), 674.
82. Wüstenhagen, R., Wolsink, M., & Bürer, M. J. (2007). Social acceptance of renewable energy innovation: An introduction to the concept. *Energy Policy, 35*(5), 2683–2691.
83. Xiao, Y., & Watson, M. (2019). Guidance on conducting a systematic literature review. *Journal of Planning Education and Research, 39*(1), 93–112.
84. Yi, J., Dai, S., Li, L., & Cheng, J. (2024). How does digital economy development affect renewable energy innovation? *Renewable and Sustainable Energy Reviews, 192*, 114221.
85. Zhang, D., Karplus, V. J., Cassisa, C., & Zhang, X. (2014). Emissions trading in China: Progress and prospects. *Energy Policy, 75*, 9–16.
86. Zhang, D., Zheng, M., Feng, G. F., & Chang, C. P. (2022). Does an environmental policy bring green innovation in renewable energy? *Renewable Energy, 195*, 1113–1124.

Chapter 5

EVALUATING RENEWABLE ENERGY POLICIES

Contents

INTRODUCTION

It is widely acknowledged that a swift transition from fossil fuels to more sustainable energy sources is necessary to prevent the harmful consequences of accelerated climate change. Investment in renewable fuel and power projects reached a significant milestone in 2013, at least USD 301 billion. This substantial investment was crucial in stabilizing global CO_2 emissions. Nevertheless, a significant increase in investment is crucial for the global transition to renewable energy as the primary energy source (Kilinc, 2016).

Various systems worldwide, such as emissions trading schemes and carbon prices, aim to indicate the economic costs of fossil fuels' complete usage to energy consumers. However, these systems often fall short in their effectiveness. Many countries have implemented policies to decrease dependence on fossil fuels and encourage the adoption of renewable energy sources. According to REN21, many countries had implemented policies to support renewable energy and set targets by early 2015. There are various methods that countries can employ to encourage the consumption and production of renewable energy for the purposes mentioned. One possible approach is to consider implementing mandates and regulations that promote the use of renewable energy sources, such as imposing bans on the use of fossil fuels (Horschig and Thrän, 2017).

Additionally, providing financial support for RE production and investing in research and development for RE technologies could be effective policy options. Renewable energy policies often have a fiscal cost; their effectiveness is frequently uncertain and can impose economic costs. There is often controversy surrounding renewable energy policy, with some commentators pointing out errors in certain policy decisions (Okwanya et al., 2021).

This chapter will explore three main areas. Firstly, the goals and policies that are implemented to achieve RE targets. Secondly, there are various arguments in favor of government support for RE policies. Lastly, a decision-support approach can assist policymakers in making informed decisions regarding renewable energy policies. Section 2 summarizes the high-level climate change and energy goals pursued by countries. Next, the focus is shifted to critically examining the justifications for RE policies, considering factors such as policy effects and outcomes and policy expenses. Section 4 considers the barriers and opportunities to implementing reformed energy policies. Section 5 highlights the various energy, economic, and geographic contexts countries should consider when implementing renewable energy policies. The key points in these sections are utilized in Section 6 to formulate energy policy principles and establish a fundamental decision-support framework for policymakers (Scott et al., 2020).

5.1. GOALS

It is important to recognize that the deployment of RE should not be seen as a standalone goal. Instead, it should be viewed as a means to contribute to specific public policy objectives, as highlighted by Edenhofer et al. in 2013. According to Dweck (1988) governments aim to balance energy security, environmental protection, and economic growth. When considering policy choices, it is important to determine the criteria to be used to evaluate the potential contributions of RE supply to various goals.

Numerous countries have set their sights on boosting the proportion of electricity produced from renewable sources, reducing reliance on fossil fuels. Rising industrial demand, rising incomes, increasing populations, or a combination of all three tend to fuel rising electricity demands in growing economies. Renewable energy is a crucial solution to address the increasing need for electricity.

However, it is challenging for users to differentiate between electricity generated from burning fossil fuels or utilizing renewable sources. There are various compelling reasons for energy policy to prioritize renewable energy (Sachs, 2012). The top nine justifications for supporting renewable energy policies are listed below:

i. As a counterbalance to the fossil fuel subsidies.
ii. To counteract the benefit that comes from fossil fuels not having To be internalized in their external costs.
iii. To achieve health co-benefits by substituting fossil fuel-generated energy
iv. To enhance the availability of low-risk energy options.
v. To create green jobs and green growth.
vi. To expedite technological advancement and establish leadership in RE.
vii. To decrease dependence on foreign fossil fuels and enhance the stability of internal energy sources.
viii. In developing countries, there is a need to enhance access to cutting-edge energy supply and minimize transmission and network expenses.
ix. To minimize energy transmission losses and enhance regional energy security (Eccles and Wigfield, 2002).

5.2. CRITIQUE OF ARGUMENTS FOR RE POLICIES

5.2.1. Subsidies for Renewable Energy as a Counterbalance for Fossil Fuels

One way to demonstrate the effects of subsidies for RE is shown in Figure 5.1. Government policies such as subsidies and taxation greatly influence energy price signals in most countries, which often lead to distortions. Additionally, the failure to account for external costs further complicates the accuracy of these signals. Numerous studies have examined and analyzed the extent and impact of subsidies for energy sources, including fossil fuels (Venetsanos et al., 2002). Various organizations, such as OECD and IEA have conducted studies to estimate the subsidies given to fossil fuel production. However, these organizations differ in the subsidy magnitudes and the methods used. One approach the OECD uses is to focus on tax expenditures and direct budgetary support that supports fossil fuel consumption or production compared to other products or activities. However, this approach overlooks the costs associated with negative externalities. In 34 OECD countries (2005–2011), the annual energy subsidies were estimated to be between $50–90 billion (OECD 2013a) (Bougette and Charlier, 2015).

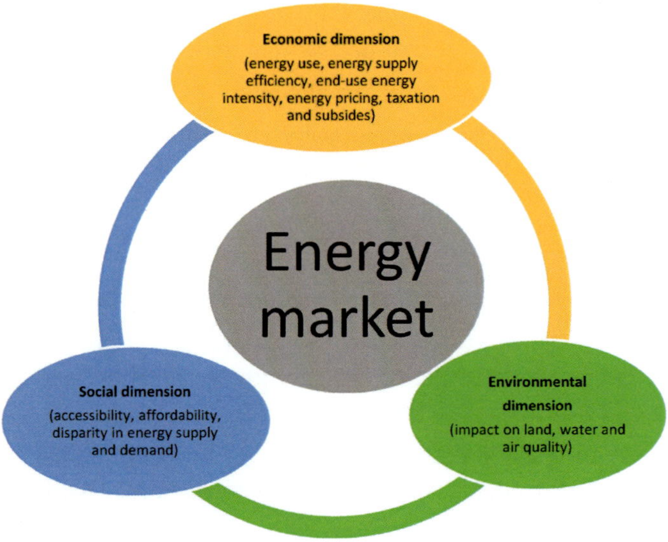

Figure 5.1. The energy market.

Source: Bluszcz A, Manowska A. (2021). The use of hierarchical agglomeration methods in assessing the polish energy market. Energies, 14(13), 3958. https://doi.org/10.3390/en14133958, Creative Commons License).

Both IMF and IEA utilize a price-gap approach, analyzing the difference between reference and end-user energy prices. The reference prices include any value-added tax, margins, total supply cost, and shipping. According to the latest estimate by the IEA, global fossil fuel subsidies reached $548 billion in 2013 (IEA). The (IMF) presents subsidy estimates using two definitions of energy subsidies. There are two types of subsidies: pre-tax subsidies and post-tax subsidies. Pre-tax subsidies look at consumer prices concerning supply cost, while post-tax subsidies consider the optimal level of taxation, supply cost, and consumer prices. The optimal level of taxation includes a consumption tax component for revenue purposes and an excise component to address externalities (Andor and Voss, 2016). Figure 5.1 illustrates the concept of external costs and subsidies. The IMF definition breaks down the subsidies into pre-tax and post-tax categories. According to their estimates, the pre-tax energy subsidies amounted to $492 billion, while the post-tax energy subsidies reached a staggering $2.0 trillion in 2011 (Day and Day, 2022).

The health effects of particle emissions and climate change are significant factors in the external costs associated with environmental policies. These costs vary greatly between countries and generating sites. However, implementing direct approaches to internalize these costs can be the most efficient way to reduce PM_{10} and CO_2 emissions (Karlilar and Balcilar, 2024).

Subsidies for fossil fuel use and production are prevalent in numerous countries, bolstering the competitive advantage of fossil fuels. One effective solution to promote RE and create a fair market is eliminating fossil fuel pre-tax subsidies. Eliminating distortions from the energy markets will result in various advantages, such as decreased reliance on fossil fuels due to their rising market prices, decreased CO_2 emissions, increased motivation to enhance fuel efficiency, and improved fiscal balance for the government. Subsidies for RE can be costly in enhancing its competitiveness, and they may not always align with other objectives like energy efficiency (Barreto, 2018).

5.2.2. Fossil Fuels and External Costs

Another argument favoring RE policy is due to the failure to account for the external costs of fossil fuel usage. This results in artificially low prices for fossil fuels in energy markets, which are below their optimal level. As a result, CO_2 emissions and fossil fuel consumption exceed their efficient levels. CO_2 emissions will have long-lasting and widespread effects on the bio-geosphere, nations, firms, and households. Without intervention, fossil fuel markets do not account for the external costs that energy consumers or producers should shoulder. As a result, fossil fuels enjoy an advantage over other energy sources that are closer to covering their costs. Some argue that RE has minimal external costs compared to fossil fuels and is less competitive in energy markets. Implementing a RE subsidy would enhance its market competitiveness, allowing it to replace some fossil fuel usage and lower CO_2 emissions (Jorli et al., 2018).

Subsidizing the RE supply is a viable strategy for reducing CO_2 emissions, as (Kalkuhl et al., 2013) suggested. Typically, direct approaches are the most cost-effective methods to redress an externality. According to Carbonell et al. (2007), implementing a tax on the carbon content of fossil fuels can lead to multiple responses in economies with various energy sources:

1. Transitioning from fossil fuels to renewable power generation.
2. Transitioning from coal to natural gas to nuclear energy in power generation.
3. Improving energy efficiency in appliances and structures to lower the electricity demand.
4. Using fewer electricity-consuming devices to reduce the electricity demand.
5. Enhancing the fuel efficiency of vehicles.
6. Encouraging a decrease in vehicle usage (Krewitt, 2002).
7. Reducing the direct fuel usage (such as natural gas) by industry and firms.

On the other hand, subsidies for RE generation will only encourage the initial response.

The impact of carbon charge policies on emission reduction is influenced by the price sensitivity of fossil fuel demand and the amount charged per tone of emissions. If both are modest, carbon charges will have a limited impact on altering energy usage. Based on extensive econometric research, it has been proven that British Columbia's carbon tax, set at C$30 per tone of CO_2e, is highly effective in reducing CO_2 emissions and fossil fuel consumption (Zhang et al., 2007).

Tradable emissions permit platforms offer a viable option to carbon prices and may be carefully crafted to cap emissions effectively. When the cap is sufficiently strict to limit emissions compared to the usual business practices, tradable permit systems offer an affordable means of reducing emissions.

Numerous studies have examined the cost-effectiveness of RE policies and have determined the expense per tone of CO_2 emissions prevented; the Australian Climate Change Authority, (OECD). In Australia, the analysis of various RE schemes revealed a wide range of costs per tone of CO_2e, from A$22 to $95. According to an ex-ante study by Le et al. (2013), the cost of avoiding CO_2 emissions through wind power varied between €56.6/t CO_2 and the maximum of €168.0/t CO_2 during the years 2007 and 2010, using a carbon price of €10 per t CO_2. According to a study by Kreps (2020), the estimated cost for the California Solar Initiative ranges from US$130 to $196 per metric ton of CO_2 emissions that are avoided. According to an ex-post study by Owen (2006), Germany had an average implicit carbon price of €57/ tCO_2 for wind power from 2006 to 2010.

In contrast, the mean implicit carbon price for solar power during the same period was much higher at €552/ tCO_2. The OECD estimates "effective carbon prices" in countries that have implemented either implicit or explicit carbon pricing schemes. They discovered convincing evidence to back up the economic claim that broad-based carbon trading schemes and taxes offer more affordable means of lowering CO_2 emissions for a variety of instruments in the transportation, energy, and household, transport, and electricity sectors, especially when they are directed at the externality (OECD).

5.2.3. External Effects Including Fossil Fuels

When coal or wood is burned, small particles (PM_{10}) are released into the atmosphere. These particles can enter the lungs and have serious health consequences, including a higher rate of illness and death. One effective way to address the health costs and other negative impacts of using petroleum products, such as road congestion, is through implementing various policies. These policies can include mandates for removing particles before emissions, regulations on the location of energy generation facilities, congestion charges, levies on polluting fuels, and the prohibition of dirty fuels. These policies directly contribute to reducing external costs associated with fossil fuels (Jorli et al., 2018). The second-best, fiscally costly option is to offer a price premium for producing low-risk and clean RE or to subsidize the installation of RE sources with the hope that they will partially replace the use of fossil fuels and the associated negative health impacts. Providing subsidies for RE may be more justifiable if fossil fuel users are not held accountable for the complete costs of consumption or generation (Rentschler and Bazilian, 2017).

It is important to be cautious as renewable electricity may not fully account for all its imposed costs. For instance, the production of hydroelectricity needs the retention of water in streams to power turbines. It is important to consider the significant ecological impacts that dams can have, such as potentially endangering certain eel and fish species and disrupting migration patterns (Hoel and Kverndokk, 1996). Wind power turbines can have noticeable impacts on the surrounding landscape and may result in financial burdens for nearby residents. Although they might not have to internalize these expenses, they may still have ecological effects, such as killing passing birds. According to Sakulniyomporn et al. (2011), biomass is second to coal regarding the externalities linked to its use in electricity generation.

According to a report by ECOFYS (2014), the weighted and range averages and the external costs for electricity technologies in EU28 countries have been documented. It is widely recognized that using fossil fuels in electricity generation incurs significant external costs. Renewable electricity technologies, excluding to a lesser extent biomass and geothermal, have significantly lower external costs than fossil fuel-powered electricity generation (Marques et al., 2018).

5.2.4. Green Jobs and Green Growth

Numerous policies implemented by governments have sectoral or local impacts on employment. Policies like renewable energy obligations, investment support schemes, and feed-in tariffs can encourage RE companies to increase production and create more job opportunities. Exploring innovative approaches to RE generation systems, evaluating their effectiveness, overseeing installation and sales, and managing their operation all positively impact economic growth and job creation (Tănasie et al., 2022). It is not clear why job creation or industry development in one area of the economy should be prioritized over job creation or industry development in another, especially if the economic activity and new jobs are merely displacing those in another area of the energy sector unless there are labor market failures (Aceleanu, 2015).

5.2.5. Accelerated Technological Progress

The government plays a crucial role in advancing clean energy technologies, as Lee (1988) emphasized. Given the nature of energy as a commodity and the challenges of price differentiation, focusing on innovative solutions that can effectively reduce production costs is crucial. It is a common challenge in energy markets to deal with price volatility, especially regarding clean energy technologies. These technologies often face difficulties in achieving grid parity, especially in cases where competing suppliers are not required to cover their entire production costs or, even worse, receive subsidies from the state. Energy infrastructure can experience technology lock-in since it is frequently long-lived and large-scale (hydropower plants might last a century or longer). Smaller players in the energy markets will face challenges when competing against well-established companies that have already recovered their start-up and fixed costs. There is a case for government intervention in the nascent industry to accelerate technological advancement. Governments can intervene on supply (carbon taxes, tax breaks, and direct grants) and demand (net metering, guaranteed market access, and feed-in tariffs (FIT)) if they are unwilling to let large incumbent companies dictate the pace of technological innovation. It could draw relatively small businesses into the clean energy innovation space. There is a high probability of spill-over effects from research and development in RE. According to Kouvaritakis et al. (2000), a compelling argument exists for government investment in basic energy research to speed up technological progress.

According to Coccia (2024), the global policy landscape has largely influenced the expansion of RE technologies. This landscape has successfully fostered economies of scale, created markets, and attracted investment, which in turn have supported technological advancements. In turn, this has resulted in decreased costs, ultimately driving long-term growth. Recent research in the United States has provided valuable insights into the government's involvement in promoting clean energy innovation. The innovation process involves various stages: development, testing, scaling, and establishing feasibility. It also entails introducing new processes or technologies, some of which can be time-consuming and capital-intensive. Established energy companies have largely controlled many of the clean energy innovation efforts. It has been observed that smaller US firms have made notable progress in the development of RE technology, surpassing the achievements of larger incumbent firms (Capros and Mantzos, 2000).

Government R&D on renewable energy technologies could more effectively accelerate technological advancement than tax credits and R&D grants for private sector firms. Similarly, which technologies possess the best prospects of future development—Solar PV is electronic and youthful, wind power is mechanical, and hydropower technology is mature—and which stage(s) of introduction, innovation, and research may be accelerated by supportive government policies should be considered by nations looking to accelerate the technological advancement in RE generation. Many countries are now prioritizing goals beyond just GDP growth. Weyant (2011) poses more visionary and larger questions regarding new industrial policy, such as what kind of nation aspires to be in 2 to 3 decades, what aspects of welfare (ecological sustainability, social objectives, and income), and what national capacities will enable competitiveness and growth in line with those pillars? For countries looking to pursue a forward-thinking industrial policy, it is crucial to prioritize sustainability, skill-rich employment opportunities, and competitiveness. RE policies can potentially drive countries towards progress on the 'high road.'

5.2.6. National Energy Security

Many countries rely on energy imports and face the potential challenges of transport disruptions and the financial burden of energy imports, including fossil fuel financing. A government should aim for greater energy independence, but some quantitative research is required before enacting laws supporting RE production on the domestic front. A national energy security index was developed by the World Energy Council in 2013, providing a useful tool for evaluating the effects of various national policies. When considering policies supporting renewable energy supply, evaluating their cost-effectiveness in enhancing national energy security is important. However, it becomes challenging to justify such policies if they are ineffective at improving energy security or come at a high cost (Bompard et al., 2017).

5.2.7. Enhanced Access to Contemporary Energy Sources

It is a staggering fact that more than one billion people are forced to rely on traditional biomass for their energy requirements since they have no access to electricity (World Bank, 2015). Off-grid, small-scale renewable energy systems have the potential to generate electricity locally, eliminating the need to wait for large-scale industrial electricity generation and distribution to reach households. Considering the potential cost-effectiveness and benefits of targeted policies, it is crucial for lower-income countries to actively pursue approaches like investment subsidies for households and communities to promote off-grid renewable energy. It can greatly enhance millions of people's access to modern energy (Bonan et al., 2017).

5.2.8. Increased Local Energy Security

One energy policy approach minimizes energy loss during transmission and enhances supply security. This can be achieved by encouraging small-scale RE systems near electricity consumption centers. Nevertheless, those potential benefits may be countered if more renewable energy is used and increases network instability in the electrical grid (Sovacool, 2013). Renewable energy sources like wave, solar, and wind power are intermittent electricity generators. Electricity storage remains a significant challenge due to its high cost and the instability it can cause in electricity systems with intermittent generation. Despite the benefits of small-scale renewable energy, the potential for greater system instability could counteract gains from a more secure local supply and reduced transmission losses. It is not advisable to adopt renewable energy policies that do not improve local energy security and, if anything, make it worse (Dobrowolski, 2021).

5.3. WHERE NEXT FOR ENERGY SUBSIDIES?

Several analysts have said that the recent decline in oil prices and the ensuing decrease in the energy portion of firms' and households' budgets presents a chance to alter taxation structures in numerous nations drastically. Energy subsidies are abundant and mostly support fossil fuels. They have a detrimental impact on energy markets, leading to increased greenhouse gas emissions and energy consumption. According to Christophers (2022), fossil fuel subsidies accounted for 36 % of global CO_2 emissions from 1980 to 2010. They hinder the adoption of emissions-free renewable energy. Eliminating fossil fuel subsidies offers a cost-effective and direct solution to address the imbalance in energy markets. This action will lead to positive outcomes such as lower carbon emissions and greater opportunities for renewable energy to thrive.

Several case studies have been reported by Van der Burg et al. (2015) and IMF (2013), highlighting instances where countries have implemented reforms in their energy policies. Numerous nations have successfully decreased or eliminated energy subsidies without causing significant harm to household incomes and business viability. Some have started by targeting higher-income households' products, such as petrol, before moving on to diesel and kerosene. Others have taken advantage of periods when prices of fossil fuels are low, such as in Indonesia. Certain countries, like Peru, India, and Angola, have implemented phased changes, while Poland and Germany have employed policies to minimize adjustment costs (Badcock and Lenzen, 2010).

Proposed guidelines for policy reform can be found in several works. Estimates of the benefits, including environmental, human health, fiscal, and economic aspects, are given by (Barkhordar et al., 2018). If post-tax energy subsidies were discontinued in 2015, it is estimated that governments would see a significant increase in revenue, equivalent to 3.6 percent of global GDP (2.9 trillion USD). Additionally, global annual emissions of CO_2 would decrease by over 20 %, leading to a substantial reduction in premature deaths caused by air pollution by more than 50 % (Fu et al., 2016).

There has been a response from international organizations and governments regarding the criticism of the significant fiscal cost associated with certain RE policies (OECD 2015). Renewable energy producers have seen a boost in supply thanks to feed-in tariffs implemented in various countries. However, consumers in Germany have experienced higher electricity prices, governments have observed the growing fiscal burden, and the need for support has decreased as production costs have dropped with advancements in wind and solar PV generation technologies. In certain countries, competitive tender systems have been implemented as a replacement for FIT, while others have opted for feed-in premiums. In 2014, the European Commission State Aid guidelines directed EU countries to implement tendering systems for renewable energy projects. By 2017, all new RE systems were required to use tenders (REN21). Over

20 countries have implemented green certificate trading systems, specifically for renewable energy. Several countries have recently implemented tax rate reductions to incentivize the growth of RE sources. Including the USA, 23 countries have established production or investment tax credits. Governments can modify tax and trading incentive programs in response to rising adoption rates, technological advancements that boost the competitiveness of RE sources, and growing awareness of the programs' financial implications (Mundaca, 2017).

Considering the various factors at play, it is crucial to assess the merits of each argument for supporting renewable energy in energy policy on a case-by-case basis in each country. It involves considering factors such as the availability of abundant RE sources, the health consequences residents face due to emissions from local coal power stations, the feasibility of attaining technological leadership in RE, and other relevant considerations (Hart, 2019). When considering the potential benefits of favoring renewable energy, it is important to evaluate various policy aspects thoroughly. It includes assessing the accuracy of policy targets, the effectiveness of the proposed policy, the necessary duration for policy implementation, and the anticipated costs involved. Only through careful evaluation can one assess the effectiveness of RE-supportive policies. When developing energy policies, it is crucial to consider the energy, trade, resource endowments, and nation's location contexts. These factors play a significant role in shaping effective policies. The following sections suggest several key tests that countries can use to guide the development of their RE policies (Bougette and Charlier, 2015).

5.4. CONTEXT

Context is crucial for countries that are contemplating the implementation of RE policies. 47 countries are island nations, accounting for 26 % of all nations. These countries are often self-reliant in generating electricity. When countries have neighboring nations, they have the opportunity to connect their electricity transmission systems. It enables a more efficient matching of electricity supply and demand, as they can share transmission and generation systems (Akaka et al., 2015).

There is variation among countries in terms of their non-renewable and renewable energy resources. Considering their location is crucial – are they situated near or far from areas where energy demand is highest? Electricity can be transmitted through cables, but energy losses are associated with this process. In the USA, these losses averaged 6.5 % in 2007, according to the US Energy Information Administration in 2009. Electricity can be transmitted over long distances and underwater using high-voltage direct current (HVDC) systems. This technology is utilized worldwide, such as between the North and South Islands of New Zealand; mainland Australia, and Tasmania; and Continental Europe and Great Britain. Currently, submarine cables can reach lengths of up to 600 km. While HVDC transmission boasts lower energy loss than AC transmission, additional costs are associated with switching between AC and HVDC (Coutaz et al., 2005).

There is a significant imbalance in the distribution of nuclear and fossil fuel resources within and among countries. Many countries rely on imports of fossil fuels or necessary elements for nuclear plants to generate electricity. It is important to note that countries vary regarding the accessibility of biofuel feedstock in convenient locations. In some cases, countries may need to import these fuels to facilitate biofuel production. Certain electricity generation facilities, like nuclear power plants, have a significant capacity for producing base-load electricity. These characteristics make them ill-suited for seamless integration or small economies as intermittent electricity providers. Wave, solar, and wind power generators have the drawback of being intermittent electricity providers and cannot be dispatched on demand. Some biofuels may have a limited supply during certain seasons, so storage or using other feedstock may be necessary to ensure a consistent power supply (Yee and Thompson, 2016).

Nations setting RE goals and implementing policies to support renewable electricity must consider electricity demand forecasts and the existing electricity generation and transmission infrastructure. Investment is crucial in areas where electricity demand is rising, and the current generation and transmission infrastructure cannot keep up. Emphasizing renewable energy investment can effectively address the need for more electricity (Smith, 1979). In areas where demand remains stable, and infrastructure is sufficient, there is significantly less urgency or potential for investing in new power generation, including renewable energy. Even when there is no increase in demand for power, investments in RE sources might assist offset the requirement to generate electricity shortly if certain infrastructure is outdated or nearing the end of its life. Considering a country like New Zealand, where RE projects have already been approved for construction to meet future demand, there may be less need for policies that encourage additional RE generation (Chalmers, 2004).

Renewable electricity can replace electricity generated from fossil fuels such as diesel, gas, and coal. It can also serve as a viable alternative to traditional fossil fuel-powered vehicles, offering a similar level of service. When deciding about technology in homes, industry, or transport, decision-makers need to consider capital and fuel costs. RE technologies like solar PV have seen remarkable reductions in capital costs over the past decade (Wolf and

Bileschi, 2006). These reductions have been driven by improved installation expertise, increased production scale, and technological advancements. For instance, depending on the location, the overall installation costs of utility-scale photovoltaic systems decreased by 29% to 65% between 2010 and 2014 (IRENA, 2015).

Long-lasting energy-consuming technologies can be heavily reliant on capital investment. At the same time, the fuel cost can fluctuate significantly, as was the case in 2014–2015 when the price of Brent Crude oil dropped by 65 %. Between October 2008 and October 2014, the gas price in North America experienced a significant decline of 39 %. This drop can be attributed to the increased supply brought by horizontal drilling and hydraulic fracking in the market, as reported by EIA in 2014. When considering the implementation of policies like feed-in tariffs (FIT) to encourage renewable energy, countries need to take into account the potential changes in capital costs for renewable energy, the possibility of lock-in and oversubscription with budget implications, and the fluctuation of costs of energy carriers like gas or oil over time (Greenberg, 2001). Decreasing gas and oil prices have the potential to quickly diminish the impact of FIT in encouraging the adoption of renewable energy. On the other hand, nations that implement subsidies to achieve equal costs between fossil fuel-generated electricity and renewable energy will face increasing financial burdens in the event of a significant drop in fossil fuel prices (Kirsh, 2001).

Regulation and legislation within countries play a crucial role in shaping the structure of the energy sector. Several countries maintain a distinct division between their retail, transmission, and generation sectors. In some other countries, companies have a vertically integrated structure where they handle both electricity retailing and generation. It is crucial for policy development to consider the specific energy sector structure in a country. When deciding on RE regulations, nations must decide which are relevant (Over et al., 2007).

Systems thinking is crucial for evaluating energy policy since implementing new rules piecemeal may have consequences contrary to the sought goals. Germany implemented a series of policies known as Energiewende in 2010, which aimed to encourage the adoption of RE sources and decrease reliance on fossil fuels, thereby reducing CO_2 emissions. The results have been shocking: traditional energy company business models were severely damaged, with significant losses in energy company value; coal use has skyrocketed in Germany; greenhouse gas emissions have increased; electricity prices have occasionally been negative. Before implementing renewable energy policies, it is necessary to model how they will affect important metrics, including economic and fiscal costs, national energy security, and global and national CO_2 emissions (Schmidt et al., 1999).

5.5. PRINCIPLES FOR RE POLICY DESIGN

Tackling climate change is a significant undertaking. According to the OECD (2013a, b), meeting the challenge will require countries to implement cost-effective policy instruments. Before modifying or developing policies opposing or supporting renewable energy, it is crucial for nations to carefully consider their geographic, economic, energy, and other relevant contexts. By asking some fundamental questions, we can decrease the chances of implementing renewable energy policies that are harmful, ineffective, or unnecessary. A methodical approach to policymaking is better suited (Howlett and Rayner, 2007).

High-income nations cite one dynamic and eight static justifications for enacting renewable energy regulations. Suppose countries determine that any or all of these nine situations are applicable to their circumstances. In that case, they may consider implementing measures to tackle energy supply or usage concerns, aiming to advance their national objectives. Here are six principles in policy design that can help inform policy choices (Punter, 2007).

1. Utilize direct approaches to address the energy problem.
2. Implement policies that effectively address the energy issue.
3. Implement cost-effective policies to address the energy issue.
4. Utilize policies that clearly understand their impact on the energy system.
5. Utilize policies with understood and known economy-wide effects.
6. Utilize policies that are practical to implement, challenging to circumvent, and have reasonable costs for compliance (Gunningham and Sinclair, 1999).

By implementing these design principles, countries can be guided toward adopting predictable, low-cost, and effective policies while avoiding the pitfalls of indirect, ineffective, expensive, and spillover-undermined, high-cost, ineffective, and indirect measures. Implementing policies such as phasing out energy subsidies and carbon pricing can encourage a range of positive reactions. Indirect policies, like subsidies for RE generation, can have varying effects and may not always produce the desired outcomes. In some cases, they may even result in carbon leakage through national borders (Gilsing et al., 2010).

Decision support systems have been created to address various environmental and resource concerns. These include the allocation of environmental resources, fisheries policy, and the selection of biodiversity reserves. Decision support approaches typically involve several key steps, such as identifying a problem that needs to be addressed, exploring different options, choosing the most suitable option, and finally implementing that option. A decision-support approach is one way to approach the decision-making process for achieving energy-related goals (Clarvis et al., 2014).

Figure 5.2 showcases a range of tests that could be applied to RE policies before implementation. By evaluating RE policies using these tests, decision-makers can determine if there is a strong enough rationale to implement a new policy. For a renewable energy policy to be considered effective, it is crucial to demonstrate that external costs and energy market distortions have been minimized to the greatest extent possible. If those criteria are satisfied, significant attention can be devoted to energy system simulation, conducting cost-effectiveness analysis, and other forms of examination of alternative policies. It will ensure that decision-makers thoroughly understand the consequences of their policy choices (Vigar and Healey, 2002).

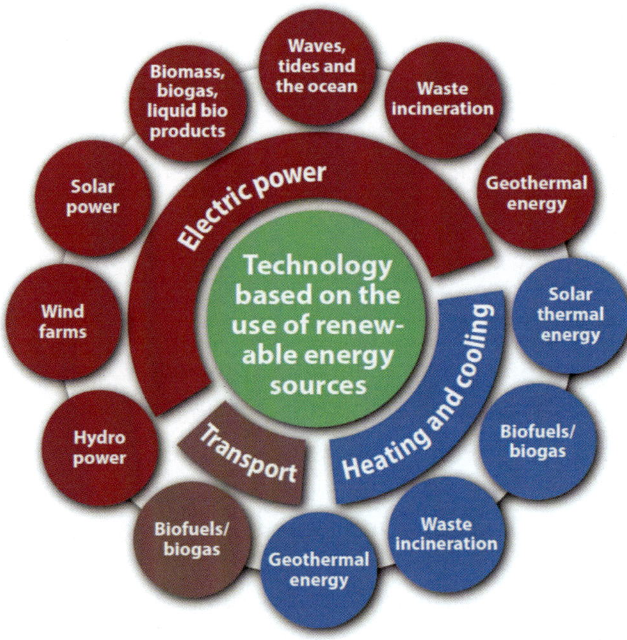

Figure 5.2. A proposed decision-support framework for renewable energy policies is shown (Source: Baskutis S, Baskutiene J, Navickas V, Bilan Y, Cieśliński W. (2021). Perspectives and problems of using renewable energy sources and implementation of local "green" initiatives: A regional assessment. Energies, 14(18):5888. https://doi.org/10.3390/en14185888, Creative Commons License).

Interestingly, several countries have opted for RE policies to address energy challenges despite the availability of more direct and cost-effective alternatives.

Reverse auctions have proven to be effective in setting FIT prices for renewable energy. They offer investor certainty and prevent the FIT rate from misaligning generation costs. It helps to avoid over-subsidization and low take-up, the issues early FIT schemes faced (Wilson et al., 2013).

5.6. SUMMARY

Renewable energy has become crucial to numerous nations' energy consumption and supply. A significant and consistent advancement in wind and solar PV power generation has resulted in increased competitiveness and reduced costs. This advancement has effectively countered the recent decline in fossil fuel prices. It is crucial to prioritize the decarburization of energy consumption in economies to effectively reduce CO_2 emissions and mitigate the pace of climate change. Numerous barriers hinder the wider adoption of renewable energy, such as the continued subsidization of fossil fuels and the failure to consider their external costs. Addressing these issues will greatly enhance the potential for greater adoption of RE.

There are many motives for energy policy to prioritize renewable energy, both dynamic and static. However, it is crucial to thoroughly evaluate policies to ensure they employ low-cost and direct measures that are time-limited and cost-effective. Many countries have made progress in reducing fossil fuel subsidies and energy policy reform. Only a handful of countries, such as New Zealand and Australia, have implemented carbon pricing mechanisms that fully account for the true costs of fossil fuel use. There is a noticeable shift in RE policy as governments take action to decrease economic benefits for generators. They are implementing competitive methods like reverse auctions to encourage the growth of renewable energy sources.

REFERENCES

1. Aceleanu, M. I. (2015). Green jobs in a green economy: Support for a sustainable development. *Progress in Industrial Ecology, an International Journal, 9*(4), 341–355.
2. Akaka, M. A., Vargo, S. L., & Schau, H. J. (2015). The context of experience. *Journal of Service Management, 26*(2), 206–223.
3. Andor, M., & Voss, A. (2016). Optimal renewable-energy promotion: Capacity subsidies vs. generation subsidies. *Resource and Energy Economics, 45*, 144–158.
4. Badcock, J., & Lenzen, M. (2010). Subsidies for electricity-generating technologies: A review. *Energy Policy, 38*(9), 5038–5047.
5. Barkhordar, Z. A., Fakouriyan, S., & Sheykhha, S. (2018). The role of energy subsidy reform in energy efficiency enhancement: Lessons learnt and future potential for Iranian industries. *Journal of Cleaner Production, 197*, 542–550.
6. Barreto, R. A. (2018). Fossil fuels, alternative energy and economic growth. *Economic Modelling, 75*, 196–220.
7. Bompard, E., Carpignano, A., Erriquez, M., Grosso, D., Pession, M., & Profumo, F. (2017). National energy security assessment in a geopolitical perspective. *Energy, 130*, 144–154.
8. Bonan, J., Pareglio, S., & Tavoni, M. (2017). Access to modern energy: A review of barriers, drivers and impacts. *Environment and Development Economics, 22*(5), 491–516.
9. Bougette, P., & Charlier, C. (2015). Renewable energy, subsidies, and the WTO: Where has the 'green' gone? *Energy Economics, 51*, 407–416.
10. Capros, P., & Mantzos, L. (2000). European Union post-Kyoto scenarios: Benefits from accelerated technology progress. *International Journal of Global Energy Issues, 14*(1–4), 204–221.
11. Carbonell, L. T., Ruiz, E. M., Gácita, M. S., Oliva, J. R., & Rivero, N. D. (2007). Assessment of the impacts on health due to the emissions of Cuban power plants that use fossil fuel oils with high content of sulfur: Estimation of external costs. *Atmospheric Environment, 41*(10), 2202–2213.
12. Chalmers, M. (2004). A historical view of context. *Computer Supported Cooperative Work (CSCW), 13*, 223–247.
13. Christophers, B. (2022). Taking renewables to market: Prospects for the after-subsidy energy transition: The 2021 Antipode RGS-IBG lecture. *Antipode, 54*(5), 1519–1544.

14. Clarvis, M. H., Allan, A., & Hannah, D. M. (2014). Water, resilience and the law: From general concepts and governance design principles to actionable mechanisms. *Environmental Science & Policy, 43*, 98–110.
15. Coccia, M. (2024). Converging artificial intelligence and quantum technologies: Accelerated growth effects in technological evolution. *Technologies, 12*(5), 66.
16. Coutaz, J., Crowley, J. L., Dobson, S., & Garlan, D. (2005). Context is key. *Communications of the ACM, 48*(3), 49–53.
17. Day, G., & Day, C. (2022). The supply-side climate policy of decreasing fossil fuel tax profiles: Can subsidized reserves induce a green paradox? *Climatic Change, 173*(3), 27.
18. Dobrowolski, Z. (2021). Energy and local safety: How the administration limits energy security. *Energies, 14*(16), 4841.
19. Dweck, C. S. (1988). Goals: An approach to motivation and achievement. *Journal of Personality and Social Psychology, 54*(1), 5–12.
20. Eccles, J. S., & Wigfield, A. (2002). Motivational beliefs, values, and goals. *Annual Review of Psychology, 53*(1), 109–132.
21. Fu, K. K., Yang, M. C., & Wood, K. L. (2016). Design principles: Literature review, analysis, and future directions. *Journal of Mechanical Design, 138*(10), 101103.
22. Gilsing, V. A., Van Burg, E., & Romme, A. G. L. (2010). Policy principles for the creation and success of corporate and academic spin-offs. *Technovation, 30*(1), 12–23.
23. Greenberg, S. (2001). Context as a dynamic construct. *Human–Computer Interaction, 16*(2–4), 257–268.
24. Gunningham, N., & Sinclair, D. (1999). Integrative regulation: A principle-based approach to environmental policy. *Law & Social Inquiry, 24*(4), 853–896.
25. Hart, R. (2019). To everything there is a season: Carbon pricing, research subsidies, and the transition to fossil-free energy. *Journal of the Association of Environmental and Resource Economists, 6*(2), 349–389.
26. Hassan, A. (2023). External energy security elements and the riskiness of clean energy stocks: A volatility analysis. *Sustainability Accounting, Management and Policy Journal, 14*(2), 396–419.
27. Hoel, M., & Kverndokk, S. (1996). Depletion of fossil fuels and the impacts of global warming. *Resource and Energy Economics, 18*(2), 115–136.
28. Horschig, T., & Thrän, D. (2017). Are decisions well supported for the

energy transition? A review on modeling approaches for renewable energy policy evaluation. *Energy, Sustainability and Society, 7*, 1–14.

29. Howlett, M., & Rayner, J. (2007). Design principles for policy mixes: Cohesion and coherence in 'new governance arrangements.' *Policy and Society, 26*(4), 1–18.
30. Jorli, M., Van Passel, S., & Sadeghi Saghdel, H. (2018). External costs from fossil electricity generation: A review of the applied impact pathway approach. *Energy & Environment, 29*(5), 635–648.
31. Karlilar Pata, S., & Balcilar, M. (2024). Decarbonizing energy: Evaluating fossil fuel displacement by renewables in OECD countries. *Environmental Science and Pollution Research, 31*(21), 31304–31313.
32. Kilinc-Ata, N. (2016). The evaluation of renewable energy policies across EU countries and US states: An econometric approach. *Energy for Sustainable Development, 31*, 83–90.
33. Kirsh, D. (2001). The context of work. *Human–Computer Interaction, 16*(2–4), 305–322.
34. Kouvaritakis, N., Criqui, P., & Thonet, C. (2000). World post-Kyoto scenarios: Benefits from accelerated technology progress. *International Journal of Global Energy Issues, 14*(1–4), 184–203.
35. Kreps, B. H. (2020). The rising costs of fossil-fuel extraction: An energy crisis that will not go away. *American Journal of Economics and Sociology, 79*(3), 695–717.
36. Krewitt, W. (2002). External costs of energy—Do the answers match the questions? Looking back at 10 years of ExternE. *Energy Policy, 30*(10), 839–848.
37. Le, L. T., van Ierland, E. C., Zhu, X., Wesseler, J., & Ngo, G. (2013). Comparing the social costs of biofuels and fossil fuels: A case study of Vietnam. *Biomass and Bioenergy, 54*, 227–238.
38. Lee, R. D. (1988). Induced population growth and induced technological progress: Their interaction in the accelerating stage. *Mathematical Population Studies, 1*(3), 265–288.
39. Marques, A. C., Fuinhas, J. A., & Pereira, D. A. (2018). Have fossil fuels been substituted by renewables? An empirical assessment for 10 European countries. *Energy Policy, 116*, 257–265.
40. Mundaca, G. (2017). Energy subsidies, public investment and endogenous growth. *Energy Policy, 110*, 693–709.
41. Okwanya, I., Alhassan, A., Migap, J. P., & Adeka, S. S. (2021). Evaluating renewable energy choices among rural communities in Nigeria: An insight

for energy policy. *International Journal of Energy Sector Management, 15*(1), 157–172.

42. Over, P., Dang, H., & Harman, D. (2007). DUC in context. *Information Processing & Management, 43*(6), 1506–1520.
43. Owen, A. D. (2006). Renewable energy: Externality costs as market barriers. *Energy Policy, 34*(5), 632–642.
44. Punter, J. (2007). Developing urban design as public policy: Best practice principles for design review and development management. *Journal of Urban Design, 12*(2), 167–202.
45. Rentschler, J., & Bazilian, M. (2017). Reforming fossil fuel subsidies: Drivers, barriers and the state of progress. *Climate Policy, 17*(7), 891–914.
46. Sachs, J. D. (2012). From millennium development goals to sustainable development goals. *The Lancet, 379*(9832), 2206–2211.
47. Sakulniyomporn, S., Kubaha, K., & Chullabodhi, C. (2011). External costs of fossil electricity generation: Health-based assessment in Thailand. *Renewable and Sustainable Energy Reviews, 15*(8), 3470–3479.
48. Schmidt, A., Beigl, M., & Gellersen, H. W. (1999). There is more to context than location. *Computers & Graphics, 23*(6), 893–901.
49. Scott, I. J., Botterud, A., Carvalho, P. M., & Silva, C. A. S. (2020). Renewable energy support policy evaluation: The role of long-term uncertainty in market modelling. *Applied Energy, 278*, 115643.
50. Smith, S. M. (1979). Remembering in and out of context. *Journal of Experimental Psychology: Human Learning and Memory, 5*(5), 460.
51. Sovacool, B. K. (2013). An international assessment of energy security performance. *Ecological Economics, 88*, 148–158.
52. Tănasie, A. V., Năstase, L. L., Vochi a, L. L., Manda, A. M., Bo oteanu, G. I., & Sitnikov, C. S. (2022). Green economy—Green jobs in the context of sustainable development. *Sustainability, 14*(8), 4796.
53. Van der Burg, E., Alais, D., & Cass, J. (2015). Audiovisual temporal recalibration occurs independently at two different time scales. *Scientific Reports, 5*(1), 14526.
54. Venetsanos, K., Angelopoulou, P., & Tsoutsos, T. (2002). Renewable energy sources project appraisal under uncertainty: The case of wind energy exploitation within a changing energy market environment. *Energy Policy, 30*(4), 293–307.
55. Vigar, G., & Healey, P. (2002). Developing environmentally respectful policy programmes: Five key principles. *Journal of Environmental Planning and Management, 45*(4), 517–532.

56. Weyant, J. P. (2011). Accelerating the development and diffusion of new energy technologies: Beyond the "valley of death." *Energy Economics, 33*(4), 674–682.
57. Wilson, D. S., Ostrom, E., & Cox, M. E. (2013). Generalizing the core design principles for the efficacy of groups. *Journal of Economic Behavior & Organization, 90*, S21–S32.
58. Wolf, L., & Bileschi, S. (2006). A critical view of context. *International Journal of Computer Vision, 69*, 251–261.
59. Yee, E., & Thompson-Schill, S. L. (2016). Putting concepts into context. *Psychonomic Bulletin & Review, 23*, 1015–1027.
60. Zhang, Q., Weili, T., Yumei, W., & Yingxu, C. (2007). External costs from electricity generation of China up to 2030 in energy and abatement scenarios. *Energy Policy, 35*(8), 4295–4304.

Chapter 6

ENVIRONMENTAL POLICY (EP) AND PREVENTIVE ENVIRONMENTAL MANAGEMENT (PEM)

Contents

INTRODUCTION

The environment plays a crucial role in supporting life on Earth. However, the increasing levels of pollution and contamination worldwide have raised significant concerns in recent times. Various types of pollution can harm the environment, such as noise, soil, water, and air pollution. By adopting cleaner production practices and technologies, industries have the potential to reduce pollution significantly. Additionally, governments play a crucial role in ensuring compliance with pollution standards by enforcing regulations and imposing penalties on non-compliant industries. Improving transportation can be achieved by encouraging the adoption of bicycles, electric vehicles, and public transport. Governments can also enforce vehicle emissions regulations. Implementing proper waste disposal systems can greatly improve waste management practices, including composting and recycling (Gasbarro et al., 2013).

Understanding the importance of preventing pollution before it happens, adopting a preventive environmental management (PEM) stance becomes crucial. There is a significant amount of research on the history of environmental policy, as evidenced by various studies however, it is concerning that many nations have not effectively utilized environmental policy to address pollution on a global scale. Numerous constraints hinder the implementation of environmental policy. In 2022, a thorough analysis was conducted to reassess the strategies used in policy allocation. A computable general equilibrium (CGE) model was utilized to review environmental policy instruments, leading to improved criteria for formulating and selecting these instruments. This practical approach has allowed for the expansion of environmental policies. In a study conducted by Cohen, the focus was on the compliance, monitoring, and enforcement of environmental policies. Di Vaio et al. (2018) conducted a study on the effective implementation of EP in China, which resulted in significant green growth. Their research also highlighted the increased demand for environmental technology and the emergence of innovative solutions due to the strict enforcement of these policies.

Pollution and environmental contamination happen when dangerous materials and pollutants are discharged into the water, soil, and air, seriously harming human health and the ecosystem. Research conducted by Li, Jin has highlighted the alarming issue of air pollution on a global scale. The severity of ozone pollution in China has gained significant attention, making it a prominent hotspot for this environmental concern worldwide (Ponce et al., 2020). It is important to note that environmental pollution and contamination can arise from both artificial and natural activities. While events like wildfires and volcanic eruptions contribute to this issue naturally, most of the problem stems from human actions such as agriculture, transportation, and industrialization. The pollution can lead to serious health consequences, such as cancer, respiratory problems, and various chronic diseases. Chemical contaminants have emerged as significant pollutants, increasing the toxicity of the air, soil, and aquatic ecosystem. It is important to consider the economic impact of environmental contamination and pollution, including the expenses of cleaning up polluted areas and addressing the negative effects on human health (Hiremath et al., 2015).

Addressing global pollution necessitates converting industrial processes into less polluting and environmentally benign systems. One way to accomplish this is by identifying possible pollution sources and implementing measures to prevent or minimize their environmental impact. Implementing minimizing waste generation in industries, reducing hazardous chemicals, and using cleaner production methods are crucial strategies to consider. Implementing these prevention strategies aims to reduce and eliminate industry pollution. Implementing environmental management systems, like ISO 14001, promotes a mindset of ongoing enhancement and guarantees that industries adhere to environmental regulations. These measures collectively work towards creating an environmentally responsible and more sustainable industrial process, fostering a better world for future generations (Lawrence, 2000).

Various innovative approaches are employed to monitor and evaluate the effects of environmental regulations on contamination control in various countries and areas. Remote sensing and satellite imaging technologies have been utilized to monitor pollution levels from space effectively. It allows for a wider view of environmental changes on a large scale and evaluates the impact of regulatory measures. Encourage the formation of international research networks that foster collaboration among experts, policymakers, and scientists from different nations. By exchanging best practices, data, and methodologies, nations can improve their collective understanding of how environmental regulations affect pollution globally (Dong et al., 2022). It also involves creating global environmental indices incorporating environmental performance indicators, regulatory frameworks, and pollution data. These indices provide a way to compare regions and nations, promoting healthy competition and motivating ongoing efforts to improve pollution management. In addition, it is possible to establish multilateral environmental agreements and collaborative approaches that enable nations to work together towards shared objectives while preserving their independence. It can be achieved by allowing nations to set their own environmental goals, considering their specific circumstances. This sharing can make sharing best practices, information, and resources for managing pollution easier. In addition, coordinating global environmental policies through platforms such as the United Nations can be instrumental in ensuring that the concerns and interests of all nations are duly considered. Strengthening institutions is crucial for effective global environmental pollution control. Humanity can make significant strides in mitigating environmental effects by improving institutional quality. It is important to focus on formulating recommendations that can be applied in real-world situations (Khan, 2017).

Amidst the multitude of environmental challenges our world faces, it is crucial to urgently tackle the significant gaps that exist on a global level to control and reduce environmental pollution effectively. Given the escalating impacts of climate change, it is crucial to emphasize the significance of proactive preventive management approaches and environmental policies. Urbanization and the rental of natural resources have been shown to affect environmental indices negatively. Taken together, these factors affect carbon footprint, ecological footprint, and carbon emissions. Another study demonstrated the correlation between CO_2 emissions and population density. They discovered a potential negative impact of biocapacity and economic growth on the nation's ecological footprint. Respecting national sovereignty is a significant obstacle when tackling global pollution. A recent study highlighted the challenge of balancing achieving effective global environmental governance and respecting national autonomy (Hermann et al., 2005).

These gaps underscore the necessity of conducting a comprehensive review to gather data on the present status of environmental pollution and to enhance the implementation and comprehension of global environmental pollution management strategies. Figure 6.1 depicts a divided planet based on its outcomes and actions. Half of the area is shrouded in a thick smog caused by industrialization, showing landscapes tainted by pollution, factories emitting smoke, and bodies of water contaminated with waste. The other half shines with the vitality of sustainable energy sources, eco-friendly urban areas, and unpolluted skies. Management tools and environmental policies are at the core of the image, representing the profound shift from pollution to sustainability. This clear contrast visually depicts the critical juncture at which global environmental pollution currently finds itself, highlighting the potential for positive change through effective environmental control policies and techniques. Understanding the laws and forces that cause environmental pollution is crucial for effectively addressing global environmental pollution. This chapter thoroughly examined and assessed the current techniques in environmental policy, pollution prevention, and environmental management systems. It also identified the gaps and emphasized the importance of taking a comprehensive approach to tackle global pollution (Chamanee et al., 2023).

Figure 6.1. Envisioning a cleaner future: from pollution to preservation.

Source: Baskutis S, Baskutiene J, Navickas V, Bilan Y, Cieśliński W. (2021). Perspectives and problems of using renewable energy sources and implementation of local "green" initiatives: a regional assessment. Energies, 14(18):5888. https://doi.org/10.3390/en14185888, Creative Commons License.

6.1. PREVENTIVE ENVIRONMENTAL MANAGEMENT

PEM employs various strategies to proactively identify and mitigate potential environmental risks, including resource conservation, pollution prevention, and waste reduction. PEM is a cost-reduction tool that helps organizations minimize liability risks by ensuring compliance. It is implemented through various strategies, including environmental monitoring systems, environmental risk assessment, green procurement, environmental performance indicators, life cycle assessment (LCA), and environmental management (Gupta, 1995). The Ecomagination program by General Electric and the zero manufacturing waste to landfills initiative by Procter and Gamble, which focuses on developing environmentally friendly services and products, are examples of common practices in environmental policy. In a study by Kazancoglu, the efficacy of PEM strategies was evaluated. The results indicated that manufacturing companies experienced a decrease in waste generation, energy consumption, and water usage. A recent study found that chemical manufacturing companies could significantly reduce their use of toxic chemicals by implementing a Pollution Prevention and Environmental Management (PEM) system (DeJong and Langford, 2002).

6.1.1. Environmental Management

Managing pollution, ecosystems, and natural resources ensures their protection and sustainability. It involves carefully planning, implementing, and monitoring strategies to minimize human actions' adverse environmental effects. Despite the obstacles, EM provides substantial advantages, including economic benefits by creating new market opportunities for sustainable products, improving resource efficiency, and reducing pollution control costs. It also offers social benefits by enhancing community resilience, quality of life, and health. Additionally, EM ensures regulatory compliance, helping organizations avoid legal penalties and fines. Several strategies are utilized in EM, such as conducting Environmental Impact Assessments (EIA), implementing sustainable resource management practices, and preventing pollution through waste minimization and cleaner production technologies practices. Additionally, environmental management systems provide a framework for managing environmental responsibilities and driving continuous improvement in environmental performance (Zeithaml and Zeithaml, 1984).

One of the key aspects of effective environmental management involves embracing principles such as participation, precaution, and sustainability. Effective environmental management is crucial for balancing present and future needs. It prioritizes proactive measures to prevent environmental damage and pollution rather than relying solely on remediation. This approach promotes the precautionary principle, which calls for taking action to prevent environmental harm even without complete scientific certainty. Furthermore, it recognizes the importance of involving all stakeholders in decision-making, including individuals, communities, industries, and government agencies. Implementing environmental management can be challenging due to various factors, such as conflicting priorities, limited resources, and insufficient data (Karagozoglu and Lindell, 2000).

6.1.2. Environmental Management Systems (EMS)

Implementing an EMS is a systematic and structured way to effectively address and mitigate environmental concerns and their effects within a company, aiming to promote sustainable development. EMS is crucial for organizations to effectively identify and manage the environmental impacts of their services, products, and activities. It aids in establishing and maintaining a sustainable strategy for managing an organization's environmental effects, adherence to environmental laws, and cost savings related to environmental hazards. Regulatory bodies regularly conduct inspections and audits to ensure organizations adhere to EMS standards and environmental regulations. Regulatory bodies play a crucial role in safeguarding the environment by enforcing EMS standards, ensuring that organizations effectively manage their environmental impact (Melnyk et al., 2003). Implementing EMS can lead to significant cost savings by enhancing operational efficiency, minimizing waste, and decreasing energy consumption. Implementing an Environmental Management System (EMS) can also assist organizations in showcasing their dedication to environmental sustainability, thereby boosting their reputation and gaining a competitive edge. Implementing EMS includes various stages, such as (i). Policy development; ii). Planning; (iii). Implementation; (iv). Monitoring; and (v). Evaluation (Hertin et al., 2008).

6.1.3. Life Cycle Assessment (LCA)

Life Cycle Assessment (LCA) is a systematic approach used to evaluate the environmental impacts of a service, process, or product throughout its entire life cycle, from the raw materials extraction to its final disposal. Considering the environmental impact at each stage of a product's life is crucial. It includes everything from extracting raw materials through manufacture, distribution, and usage to end-of-life management. It thoroughly explains how activities and goods affect the environment, enabling businesses and decision-makers to make sustainable decisions and minimize their ecological footprint. It is important to derive insights from Life Cycle Assessment (LCA) to develop environmental regulations and policies, environmental product declarations, and eco-labeling. The standard procedures in an LCA include life cycle inventory, decision-making, improvement and interpretation, life cycle impact assessment, and goal formulation and scope (Roy et al., 2009).

6.1.4. Environmental Risk Assessment

ERA is a structured process used to evaluate the potential negative effects of human actions on the environment. It assesses, examines, and appraises potential hazards linked to particular substances or actions to support well-informed decision-making and reduce environmental harm. Our main focus is to assess the likelihood and impact of negative consequences on different ecosystems, including atmospheric, aquatic, and terrestrial systems. ERA is widely used in different areas, such as pollution incidents, genetically modified organisms, chemical substances, infrastructure development, and industrial activities. It protects human and ecosystem well-being, directs environmental management practices, and promotes sustainable development. The basic process of conducting an Environmental Risk Assessment (ERA) includes identifying hazards, assessing exposure, evaluating effects, and managing risks (Van der Oost et al., 2003).

6.1.5. Environmental Performance Indicators

Environmental performance indicators (EPIs) are utilized to gauge and evaluate the environmental performance of various entities such as services, products, processes, or organizations. They offer a comprehensive range of data, allowing for evaluating, tracking, and documenting sustainability initiatives, resource utilization, and environmental effects. EPIs are essential for effectively communicating environmental performance to stakeholders, pinpointing areas for improvement, setting objectives, and monitoring progress. Various environmental performance indicators can be used, depending on the particular objectives and context (Azzone et al., 1996). Several commonly used environmental performance indicators (EPIs) are important to consider, including certifications, environmental expenditure, eco-efficiency, environmental compliance, biodiversity impact, recycling rate, waste generation, water usage, greenhouse gas emissions, and energy consumption (Henri and Journeault, 2008).

6.1.6. Green Procurement

Green procurement, also called environmentally responsible or sustainable procurement, entails incorporating environmental considerations into the purchasing decisions made by organizations. Its goal is to reduce the environmental impact of acquiring services and goods by considering factors like eco-friendly technologies or materials, pollution prevention, waste reduction, resource conservation, and energy efficiency (Michelsen and de Boer, 2009). Effectively implementing green procurement necessitates the cooperation of various stakeholders, such as top management, suppliers, sustainability teams, and procurement departments. It requires seamlessly integrating environmental considerations in supplier selection criteria, processes, and procurement policies along with continuous environmental performance evaluation and monitoring. Green procurement encompasses several important strategies and principles, such as fostering collaboration with suppliers, utilizing labels and product certification, conducting life cycle assessments, evaluating suppliers, and integrating environmental criteria (Appolloni et al., 2014).

6.1.7. Environmental Policy (EP)

Environmental policy involves implementing measures by governments and organizations to tackle environmental issues, exemplified by the Clean Air Act in the US. The challenges are diverse and encompass various concerns, including water and air pollution and climate change. Environmental policy has a rich history that can be traced back to the early conservation movement of the 19th century. In the United States, establishing national parks in the late 19th century was the first notable environmental program (Clarkson et al., 2008). Environmental policy saw a significant increase in the twentieth century, thanks to the establishment of the United States Environmental Protection Agency (EPA) in 1970 and the implementation of the Clean Water Act and Clean Air Act. International organizations, like the United Nations, have had a major impact on the development of environmental policies [64]. In 1972, the Conference on Human Environment took place in Stockholm, marking a significant milestone in the global advancement of environmental policy. The conference resulted in the establishment of the United Nations Environment Program (UNEP), which significantly impacted the development of global environmental policy (Liefferink and Andersen, 1998).

6.1.8. Contemporary Global Environmental Policies

6.2.8.1. International Environmental Policies

International environmental policies are crucial in safeguarding the planet and addressing environmental challenges. These policies are crucial in preserving natural resources, promoting sustainable development, and addressing global pollution. These policies address a wide range of environmental concerns, such as pollution, climate change's impact, and biodiversity loss. Countries establish regulations through international agreements and conventions, devise strategies to address environmental threats, and work together to set goals (Almer and Winkler, 2017).

6.1.8.2. United Nations Framework Convention on Climate Change (UNFCCC)

Adopted in 1992 and ratified by almost all nations, the UNFCCC is a prime example of a major international environmental policy to combat climate change. It offers a strong platform for international cooperation to stabilize GHG concentrations in the atmosphere and avoid damaging human intervention with the delicate climate system that sustains life. By adhering to the principles of the UNFCCC, nations come together to work towards a sustainable and resilient future. It involves preserving ecosystems for generations to come, protecting societies, and minimizing climate change's effects (Scott, 2015).

6.1.8.3. Kyoto Protocol

The Kyoto Protocol, a significant achievement in global endeavors to address the imminent threat of climate change, emerged due to the UNFCCC. Implemented in 1997 and fully effective in 2005, the protocol promotes compulsory emissions reduction by the developed countries from 2025 to 2012, referred to as Annex I parties. The ultimate goal is to significantly reduce greenhouse gas emissions, well below the levels of 1990, to create a future that is both sustainable and responsible. The protocol includes market-based cutting-edge emission trading systems, which drive reduction efforts and provide incentives for adopting greener practices. Despite the achievements of this protocol in raising awareness and setting the stage for subsequent international frameworks, it encounters various obstacles in its implementation (Iwata and Okada, 2014).

6.1.8.4. Paris Agreement

The Paris Agreement, adopted under the UNFCCC in 2015, is a major milestone in combating climate change. The objective is to enhance the worldwide effort and restrict the increase in global temperatures to less than 2°C over pre-industrial levels, with the ideal goal of keeping it at 1.5°C. The agreement requires all nations to make commitments to intensify efforts to support developing countries, adapt to the impacts of climate change, and reduce greenhouse gas emissions. This agreement has gained significant acceptance, but certain nations are encountering challenges in fully complying with it due to the requirement of regular reviews to assess a country's progress in achieving its climate-related objectives (Schleussner et al., 2016).

6.1.9. National Environmental Policies

Many countries have implemented national environmental policies to address pollution and minimize its negative environmental effects. These policies are customized for each country, considering their specific environmental priorities and trials. These tactics could include programs, laws, and regulations protecting our natural resources, reducing pollution, and supporting sustainable development. These policies play a crucial role in ensuring the protection of biodiversity and ecosystems and the well-being of communities (Hanks and Hanks, 1969).

6.1.9.1. United States Environmental Protection Agency (USEPA)

The USEPA, also known as the EPA, supports and implements a wide range of environmental policies throughout the United States. The primary objective is to address and protect the environment, safeguard human health, and reduce pollution. The EPA always creates and enforces many standards and regulations covering anything from hazardous waste disposal methods to chemical usage to water and air quality. The numerous initiatives the EPA undertakes are crucial in guiding the United States toward a future characterized by improved cleanliness, better health, and enhanced sustainability (McCuin and Clancy, 2003).

6.1.9.2. Environmental Policy of the EU

Embracing a more sustainable approach, the EU is prioritizing the shift towards a circular economy, moving away from the traditional linear economic model of "take-make-dispose." The concept focuses on preserving resources, reducing waste, and optimizing materials and product value. In addition, the EU is firmly committed to reducing GHG emissions and addressing climate change. By implementing stringent regulations and ambitious targets, the EU fosters the development of more efficient and cleaner energy systems, reduces reliance on fossil fuels, and promotes the adoption of renewable energy sources (Steinebach and Knill, 2017).

This commitment includes research money and financial incentives to encourage sustainable technologies. Europe's environmental policy framework is inclusive and strong, clearly focusing on reducing pollution and promoting sustainable growth. It encompasses various regulations, legislations, and strategies that tackle diverse environmental challenges and dilemmas. These strategies focus on key sectors such as climate change mitigation, biodiversity protection, waste management, and water and air quality (Liefferink and Andersen, 1998).

6.1.9.3. Environmental Protection Policies of China

As a leading global economy, China recognizes the utmost significance of effectively tackling the current environmental challenges. China has implemented comprehensive environmental protection policies to tackle pressing issues such as ecological restoration, energy preservation, and water and air pollution control. The country has implemented numerous initiatives to reduce vehicle and industry emissions greatly, encourage the use of renewable and clean energy, and significantly improve waste management practices. China is making significant progress towards a more environmentally friendly and sustainable future through a commitment to rigorous environmental impact assessments, active public engagement, and substantial investments in research, cutting-edge technologies, and essential infrastructure (Zhang and Wen, 2008).

6.1.10. Regional Environmental Policies

Regional environmental strategies play a crucial role in reducing pollution in specific geographical areas by bringing together efforts and establishing unified frameworks for environmental protection among collaborating nations. It enables a more efficient response to the environmental challenges prevalent in the region. These strategies frequently lead to regional norms, statutes, and accords focused on reducing pollution, managing waste, preserving resources, and protecting biodiversity. These policies serve as a foundation for countries to unite, promoting shared responsibility for safeguarding the environment and fostering sustainable development in their regions (Fell and Maniloff, 2018).

6.1.10.1. Environmental Policy of the African Union (AU)

The AU's environmental policy is a comprehensive plan to tackle many pressing environmental issues throughout Africa. This policy is a promising lighthouse for Africa's environmental potential, with an uncompromising commitment to reducing pollution, protecting valuable natural wealth, and sustainable growth. The urgent requirement for strong governance, investment in human capital, and regional collaboration is highlighted to address these environmental challenges effectively. To achieve the desired impact of its Environmental Policy, the AU works closely with its member states, rigorously enforces this crucial blueprint, and strengthens it through a wide range of initiatives (Chambiwa et al., 2022).

6.1.10.2. Environmental Policy of the Association of Southeast Asian Nations (ASEAN)

The ASEAN Environmental Policy was developed to promote sustainable development and safeguard the environment in its member states. I am dedicated to positively impacting environmental issues. The policy emphasizes addressing pollution through a cooperative and complete approach. ASEAN actively works to bring about significant change by concentrating heavily on important areas such as climate change adaptation, biodiversity conservation, efficient waste management, and water and air pollution regulation. Through collective efforts, ASEAN states work together to implement the policy by exchanging best practices, collaborating on research, and establishing regional guiding policies and frameworks (Elliott, 2000).

6.1.11. Corporate Environmental Policies

Company environmental strategies are crucial in promoting pollution prevention by establishing objectives and guidelines. These strategies encourage businesses to reduce their ecological footprint by implementing PEM. These directives highlight a commitment to emission reduction, resource conservation, and sustainable practices. It provides strategies for increasing energy efficiency, supporting renewable energy sources, and reducing waste generation. Various companies, including Walmart, Apple, and Toyota, have implemented strategies to promote a cleaner environment and mitigate the negative impacts of their operations (Fernando et al., 2017).

Toyota's Environmental Challenge 2050 presents a comprehensive plan to address environmental issues associated with manufacturing and automobiles. Their policy outlines six essential objectives: fostering a sustainable society, protecting biodiversity, prioritizing alternative fuel vehicles, promoting recycling, minimizing water usage, and reducing CO_2 emissions. Toyota is committed to reducing vehicle emissions by focusing on hydrogen fuel cells and electric vehicle R&D. They envision a future with no CO_2 emissions. Apple's commitment to environmental responsibility is evident in its strong focus on reducing its carbon footprint and preserving vital resources for the planet's well-being. This policy is focused on innovative product design, promoting energy efficiency, and ensuring ethical sourcing methods. With their commitment to sustainability, Apple seamlessly integrates renewable energy sources and energy-efficient, cutting-edge technologies into their exceptional product lineup and daily operations (Sarkar, 2008). Apple is committed to supporting recycling programs and reducing waste, encouraging the international community to work together for a sustainable future. Through ensuring responsible sourcing, Apple aims to eliminate hazardous substances and significantly reduce their environmental impact across its complex supply chain [76]. As reflected in its policy, Walmart's strategy for sustainability demonstrates a strong dedication to environmental stewardship, taking a comprehensive view of sustainability. Their commitment to achieving zero waste, harnessing renewable energy sources, and revolutionizing the market with a wide range of environmentally friendly products showcases their dedication. Walmart is committed to ambitious goals, including reducing greenhouse gas emissions, improving energy efficiency, and incorporating renewable energy within its operations (Tatoglu et al., 2015).

6.1.12. Gaps in Modern Environmental Policies

Despite the wide range of environmental policies in place, significant loopholes still undermine efforts to hinder sustainable environmental progress and prevent pollution. These gaps require immediate attention due to ineffective enforcement strategies, insufficient funding for policy implementation, inadequate reporting and surveillance systems, and a lack of global coordination and unity (Busscher et al., 2014). One notable weakness is the lack of strong enforcement measures, resulting in a failure to implement outlined practices due to lenient execution and minimal consequences for non-compliance. Not only do corporations and individuals have the freedom to ignore environmental statutes, but this also results in increased pollution levels. Enhancing enforcement strategies is vital for guaranteeing compliance and the effectiveness of environmental policies. In addition, the lack of adequate funding presents a significant challenge to supporting research on reducing pollution and developing sustainable infrastructure, hampering efforts to enforce regulations. To bridge this divide, securing additional funding and allocating resources more effectively is crucial. Insufficient reporting and monitoring systems contribute to the difficulty of identifying pollution sources and implementing focused strategies. Advocating for transparency, establishing strong reporting systems, and implementing advanced monitoring technologies are crucial measures to enhance the effectiveness of policies (Dedić et al., 2020). In addition, the lack of international coordination and cooperation, caused by differences in priorities and regulations, presents challenges in addressing pollution on a global scale. To address this gap, it is crucial to strengthen international consolidation mechanisms, establish common goals, and foster dialogue among nations. These steps are necessary to ensure a unified, efficient approach to pollution prevention. By increasing investment in research and innovation, encouraging sustainable business practices, improving national environmental regulations, and implementing stronger international agreements, we can enhance the effectiveness of environmental policy in reducing environmental pollution (Panwar et al., 2014).

6.1.13. Air Pollution

Global health is at risk due to air pollution, ranging from the smog lingering over cities to the smoke infiltrating homes. Based on a recent report by the World Health Organization (WHO) on global air pollution, it is concerning that nearly the entire global population (99%) is exposed to air pollution levels that significantly raise the risk of various diseases, such as and pneumonia, cancer, chronic obstructive pulmonary disease, stroke, and heart disease. Monitoring exposure levels and health impacts is a crucial task undertaken by the WHO. They focus on tracking deaths and Disability-Adjusted Life Years (DALYs) caused by air pollution, whether it originates from outdoor or household sources, on a global, regional, or national scale. These estimates are of utmost importance in official reporting, such as the SDGs and the World

Health Statistics (Glencross et al., 2020). The air pollution data portal offered by the WHO presents concerning statistics, highlighting the significant impact of household and ambient air pollution on public health. It reveals that 6.7 million deaths occur each year due to exposure to air pollution. Furthermore, there is a pressing issue regarding household exposure, as a staggering 2.3 billion individuals still depend on polluting technologies and fuels for cooking as of 2021. Implementing various strategies, including enforcement, legislation, education, urban planning, and air quality monitoring, is crucial to effectively tackle the severe and pervasive effects of air pollution on public health globally (Bernstein et al., 2004).

6.1.14. Global Carbon Project

Assessing anthropogenic CO_2 emissions and their distribution among the terrestrial, oceanic, and atmospheric biosphere is crucial for understanding the global carbon cycle, developing effective climate policies, and predicting future climate change. According to the latest findings from the global carbon budget, the data in Table 6.1 reveals that global fossil CO_2 emissions, including cement carbonation, slightly surpassed the levels seen before the COVID-19 pandemic. These emissions reached 10.0 gigatons of carbon per year (GtC yr−1) in 2022. China saw a slight decrease of 0.9% in emissions, while the European Union experienced a reduction of 0.8%. On the other hand, India, the US, and the rest of the world witnessed increases of 6%, 1.5%, and 1.7% respectively (Friedlingstein et al., 2022). The carbon project report highlighted different trends in emissions from coal, oil, and gas, oil, and coal with changes of -0.2%, 2.2%, and 1.0% respectively. During 2012–2021, it is worth mentioning that CO_2 emissions by 24 countries experienced a decrease, making a significant contribution of about 2.4 GtC yr^{-1}. Deforestation continues to be a significant factor, as emissions reached 1.8 ± 0.4 GtC yr^{-1}. Worldwide emissions reduction initiatives must be consistent with climate targets. In 2022, the concentration of atmospheric CO_2 saw a significant 51% increase compared to pre-industrial levels (Le Quéré et al., 2013).

Table 6.1. Recent Literature Review.

Authors	**Year**	**Country**	**Methodology**	**Empirical Findings**
Urbanization-CO_2 emissions				
Le Quéré, et al., (2013)	1960–2018	102 less-developed nations	Fixed effects model	URB ↓ CO_2 emissions

Liefferink, & Andersen, (1998)	1997–2019	China	Spatial decomposition	URB ↑ CO_2 emissions
Liu, et al., (2015)	1996–2018	China	Dynamic threshold panel approach	URB ↑ CO_2 emissions
Mashima, & Okuyama, (2015)	1990–2022	China	Granger causality test	URB ↑ CO_2 emissions
McCuin, & Clancy, (2003)	1995–2020	Chinese provinces	Panel PMG-ARDL	URB does not affect CO_2 emissions
Melnyk, Sroufe, & Calantone, (2003)	1990–2019	Sub-Saharan African countries	Panel quantile regression; EKC	URB ↑ CO_2 emissions
GDP-CO_2 emissions				
Meybeck, & Helmer, (1989)	1820Q1–2021Q4	G7 countries	Panel quantile regression	GDP ↑ CO_2 emissions
Miettinen, & Khan, (2022)	1990–2014	E7 countries	Granger causality test	GDP ↑ CO_2 emissions;
Ponce, et al., (2020)	1992–2018	Russia	Quantile on quantile regressions	GDP ↑ CO_2 emissions
Schleussner, et al., (2016)	1970Q1–2019Q4	G7 countries	VAR model	GDP → one-way causal link with CO_2 emissions
Schwar-zenbach, et al., (2010)	1980–2022	Middle East countries	Heterogeneity and Westerlund cointegration test	GDP ↑ CO_2 emissions
Sigmund, et al., (2023)	1990–2020	Thailand	ARDL bound test; DOLS	GDP ↑ CO_2 emissions
GDP^2-CO_2 emissions (EKC hypothesis)				
[92]	1962–2018	China	ARDL, EKC	U-shaped EKC
[93]	1990–2019	Bangladesh	ARDL, EKC	Inverted U-shaped EKC

[89]	1971–2014	Thailand	Vector error correction model (VECM), EKC	The validity of EKC
[94]	1960–2015	Turkey	ARDL bound test, EKC	The validity of EKC
[95]	1971–2014	India, Pakistan, Bangladesh	VECM, EKC, ARDL	U-shaped EKC
[96]	1996Q1–2016Q4	India, Pakistan, Bangladesh	ARDL, EKC	Inverted U-shaped EKC
Renewable energy consumption-CO_2 emissions				
[62]	2000–2018	33 OECD countries	The panel smooth transition regression (PSTR)	RNEW ↓ CO_2 emissions
[63]	1995–2018	MINT countries	MMQR, EKC	RNEW ↓ CO_2 emissions; The validity of EKC
[65]	1990–2014	Latin American and Caribbean countries	Panel ARDL	RNEW ↓ CO_2 emissions in the short and long run
[66]	1995–2018	ASEAN nations	Panel ARDL	RNEW ↓ CO_2 emissions in the short but insignificant effect in the long run
[68]	1995–2015	120 global countries	FMOLS, DOLS, EKC	No significant effect between RNEW and CO_2 emissions
[69]	1990–2014	Sub-Saharan nations	Augmented mean group (AMG)	No significant effect between RNEW and CO_2 emissions
[70]	2018–2019 survey	Pakistan	STIRPAT model	RNEW ↓ CO_2 emissions

ICT technology-CO_2 emissions				
[71]	2006–2017	China's provincial panel data	GMM estimation method	ICT ↑ CO_2 emissions
[72]	2010–2020	E7 countries	Gray relational analysis (GRA), GMM, EKC	ICT ↑ CO_2 emissions; U-shaped EKC
[73]	2000–2018	36 OECD countries	AMG and GMM	ICT ↓ CO_2 emissions
[74]	2001–2014	EU countries	Pooled mean group (PMG)	ICT ↑ CO_2 emissions
[76]	2007–2019	China	GMM	ICT ↓ CO_2 emissions
Agriculture-CO_2 emissions				
[79]	1990–2018	South Asian countries	FMOLS, EKC	AGR ↓ CO_2 emissions
[80]	2004–2020	15 developing countries	Panel quantile regression	AGR ↑ CO_2 emissions
[81]	1971–2014	Saudi Arabia	ARDL, EKC	AGR ↓ CO_2 emissions, inverted U-shaped EKC
[82]	1971–2016	BRIC countries	Fourier cointegration and causality test	AGR ↑ CO_2 emissions
[83]	1990–2019	Malaysia	ARDL, DOLS, Granger causality test	AGR ↑ CO_2 emissions
[84]	1990–2016	China	ARDL bound test; FMOLS, CCR	AGR ↓ CO_2 emissions in the long run
FDI-CO_2 emissions				
[97]	1981Q1–2021Q1	APEC economies	Common correlated effect mean group	FDI ↑ CO_2 emissions
[98]	1980–2014	17 Asian nations	FMOLS	FDI ↑ CO_2 emissions
[99]	2000–2014	32 Asian nations	EKC	FDI ↑ CO_2 emissions
[100]	2001–2019	Asian economies	2SLS, GLS, GMM	FDI ↑ CO_2 emissions

[101]	1996–2016	40 Asian countries	GMM	FDI ↓ CO_2 emissions

Note: Arrows in the Table Indicate Increases and Decreases.

Source: Sobirov Y, Makhmudov S, Saibniyazov M, Tukhtamurodov A, Saidmamatov O, Marty P. Investigating the Impact of Multiple Factors on CO_2 Emissions: Insights from Quantile Analysis. Sustainability. 2024; 16(6):2243. https://doi.org/10.3390/su16062243.

6.1.15. Gas Emissions

According to the Emissions Gap Report 2023, published by the UNEP, there are increasing difficulties caused by GHG emissions, rising temperatures, and worsening climate effects. This report evaluates the disparity between promised reductions in greenhouse gas (GHG) emissions and the necessary reductions following the long-term temperature objective of the Paris Agreement (Johnson et al., 2007). GHG emissions reached an unprecedented high of 57.4 $GtCO_2e$ in 2022, with the burning of fossil fuels and industrial activities accounting for a substantial two-thirds of the total. The rapid increases in fluorinated gases (F-gases), nitrous oxide (N_2O), and methane (CH_4) emissions have further compounded the environmental challenge. The CO_2 emissions from global net land use, land-use change, and forestry (LULUCF) have shown no significant changes. Considering the current policies of the Paris Agreement, there has been significant progress in reducing the expected rise of GHG emissions by 2030 from 16% to 3%.

Table 6.2. *Existing Standards Associated with the Reduction of CO_2 Emissions in Buildings as Part of Prioritizing Action on the Nationally Determined Contribution (NDC), Which Was Set up in 2015 Under the Paris Agreement Commitment and the United Nations Sustainable Development Goals*

Country	Standards or Policies
China	The Energy Consumption of Buildings standard was introduced by the Ministry of Housing and Urban-Rural Development in 2016. This standard covers indicators of energy use for various building types. It aims to limit the amount of building sector energy consumption of the country and simultaneously limit the total CO_2 emissions.

Australia	The Australian Federal Government launched the National Carbon Offset Standard for Buildings in 2017. It was established in collaboration with the Green Building Council Australia. The main objective of the standard is to provide to measure, reduce, offset, report, and audit CO_2 emissions from building operations.
India	A policy was introduced in 2016 as part of the Energy Conservation Act of 2001, which is aimed at commercial buildings under the Perform, Achieve, and Trade (PAT) program. It had saved about 9 million tons of oil equivalent (Mtoe) of final energy, thereby reducing annual CO_2 emissions by nearly 23 $MtCO_2$. The Energy Conservation Building Code (ECBC) was updated in 2017 for commercial buildings that recognized improvement efforts towards decarbonization. In 2018, the first national model building energy code called the Energy Conservation Building Code for Residential Buildings was introduced with simple enforcement of thermal comfort and passive system improvement.
European Union	As part of the Clean Energy for all European policy package set in 2016, the European Commission targets to combat climate change contributed by GHGs, including CO_2 emissions, through proposals on energy efficiency, energy market, and renewable energy strategies. An amendment to the Energy Performance of Buildings Directive (EPBD) was published in 2018 to achieve high-energy efficiency and be decarbonized by 2050.
Sweden	In 2016, the Centre for Sustainable Construction was created under a Swedish Government policy to promote the usage of sustainable materials and energy-efficient renovations, which could also reduce CO_2 emissions. In 2019, a certification scheme addressing the environmental impact of a new building was introduced.
Japan	In 2017, the act on the Improvement of Energy Consumption Performance of Buildings (Building Energy Efficiency Act) was introduced, which includes regulatory measures for mandatory compliance with energy efficiency standards for non-residential buildings. This act is part of the Japanese government policy on the zero-energy-building [ZEB]/zero-energy-house [ZEH] system to be achieved by 2030.

Canada	In 2016, tighter energy performance standards for energy-using product categories in buildings were introduced. New building energy codes have been planned to be introduced in 2022 as part of the Pan-Canadian Framework on Clean Growth and Climate Change to increase energy efficiency in existing buildings. In 2019, the Government of Canada was working to develop a net-zero-energy-ready building code to support the aim of implementing building energy use labeling.
Germany	In 2019, a package of emission mitigation measures by the German government was formed to meet the requirement of Agenda 2030 in the building sector.
USA	In 2018, the California 2019 Building Energy Efficiency Standards was introduced as the first code in the USA. The New York State Energy Research and Development Authority was established in 2018 to increase the overall efficiency and sustainability of buildings.
Nigeria	In 2017, the first building energy code was introduced with a partnership between the German Development Agency (GIZ) and the Nigerian Energy Support Program with the aim of a set of minimum standards for energy-efficient building construction in Nigeria.
Singapore	In 2016, the Code on Environmental Sustainability Measures for Existing Buildings was launched in 2016 for existing non-residential buildings within Singapore's Building Control Regulations.
Switzerland	In 2018, Switzerland's new Energy Act came into force, intending to increase energy efficiency in buildings towards decarbonization. Including the use of a CO_2 tax on stationary fuels (heating and industry). Under the Act, CO_2 tax and subsidizing of geothermal energy have been introduced. A Federal Act on Reduction of CO_2 Emissions was revised in 2019 to implement NDC in the building sector.

Source: Ahmed Ali K, Ahmad MI, Yusup Y. (2020). Issues, impacts, and mitigations of carbon dioxide emissions in the building sector. Sustainability, 12(18):7427. https://doi.org/10.3390/su12187427

It emphasizes the urgent need to address disparities in emissions contribution. Accelerated and immediate mitigation actions are necessary to achieve the deep annual emission cuts required to narrow the emissions gap. One way to achieve this is by encouraging nations to speed up their efforts in transitioning to low-carbon development. It is also important for industrialized nations to take more ambitious actions while supporting developing nations as they strive for low-emission growth (Carlson et al., 2017) (Table 6.2).

6.1.16. Water Pollution

Throughout 2021, the persistent global water crisis continued to impact more than 2 billion individuals residing in water-stressed nations. Unfortunately, this situation is projected to deteriorate further due to population expansion and climate change. It is deeply concerning that in 2022, a staggering number of individuals globally were exposed to drinking water contaminated with feces, leading to significant public health threats. Microbial contamination from various sources poses a significant risk to water safety, spreading diseases such as polio, typhoid, dysentery, cholera, and diarrhea (Dwivedi, 2017). It leads to a staggering 505,000 deaths caused by diarrheal diseases each year. Despite significant progress in providing access to safe drinking water services, disparities in access still exist among the global population. The other 2.2 billion people did not have access to these services and had to deal with everything from gathering water from unsecured sources to basic access that required a 30-minute round trip. The data presented in Table 6.3 highlight the significant impact of drinking water on health worldwide. Nations worldwide need to adopt the frameworks and regulations provided by WHO to manage the risks of water pollution. The WHO's focus on health-based impacts and its guidelines for water quality are crucial in addressing this issue globally (Azizullah et al., 2011).

Table 6.3. *Most Crucial Natural Factors/Processes that Affect Water Quality*

Water Resources	Processes/Factors	Important Processes
Surface water	Hydrological process	Evaporation, suspension, and setting
Groundwater		Transpiration, infiltration, and leaching
All water resources		Dilution
All water resources	Physical process	Adsorption and desorption, diffusion
Mainly river and lakes		Heating and cooling, vitalization, gas exchange with the atmosphere
Groundwater	Chemical process	Ionic exchange
All water resources		Acid-base reactions, redox reactions, Precipitation of minerals, photo degradation, Dissolution of particles

Surface water	Biological process	Primary production
All water resources		Microbial die-off and growth
Mainly rivers and		Bioaccumulation, decomposition of organic matter, biomagnifications

Source: Akhtar N, Syakir Ishak MI, Bhawani SA, Umar K. (2021). Various natural and anthropogenic factors responsible for water quality degradation: A review. Water, 13(19), 2660. https://doi.org/10.3390/w13192660

6.1.17. Global Pollution

According to the Lancet Commission on Pollution and Health, the impact of pollution on global health is alarming. In 2015 alone, pollution was responsible for a shocking 9 million premature deaths, making it the leading environmental risk for early mortality and disease. According to the recent findings in the Global Burden of Diseases, Injuries, and Risk Factors Study 2019, pollution remains a significant cause of around 9 million deaths yearly. An example is presented in Table 6.4, highlighting that pollution accounts for one in six global fatalities (Schwarzenbach et al., 2010). Despite some improvements in reducing deaths caused by certain forms of pollution related to extreme poverty, the overall progress is overshadowed by the increasing number of deaths caused by toxic chemical pollution and ambient air pollution, specifically lead. Since 2015, there has been a 7% increase in mortality from these contemporary environmental risk factors due to the unintended consequences of urbanization and industrialization, and this number has increased dramatically by 66% since 2000 (Meybeck and Helmer, 1989).

Table 6.4. *Incidence Rate Ratio (IRR) of Mortality Associated with Each Pollutant Standards Index (PSI) Range Specified by the National Environment Agency and Each 10-Unit Increment in the Daily Average PSI Within Each PSI Range*

	Risk Associated with Each PSI Range Compared to Good PSI		Risk Associated with Each 10-Unit Increase in PSI	
	Unadjusted IRR (95% CI)	**Adjusted IRR (95% CI) ^**	**Unadjusted IRR (95% CI)**	**Adjusted IRR (95% CI) ^**
Overall	Not applicable	Not applicable	1.01	1.01
			(1.01–1.02) *	(1.00–1.01) *

Good PSI	1.00	1.00	1.02	1.01
	(reference)	(reference)	(1.01–1.03) *	(1.00–1.02) *
Moderate PSI	1.06	1.05	0.98	0.98
	(1.04–1.08) *	(1.03–1.07) *	(0.95–1.02)	(0.94–1.02)
Unhealthy PSI	1.10	1.08	0.99	0.99
	(1.05–1.16) *	(1.03–1.14) *	(0.87–1.14)	(0.85–1.16)

** statistical significance with $p < 0.05$. ^ adjusted for daily average temperature, daily average wind speed, and daily total rainfall.*

Source: Ho AFW, Zheng H, Cheong KH, En WL, Pek PP, Zhao X, Morgan GG, Earnest A, Tan BYQ, Ng YY, et al. (2020). The Relationship Between Air Pollution and All-Cause Mortality in Singapore. Atmosphere, 11(1):9. https://doi.org/10.3390/atmos11010009.

6.2. HOLISTIC APPROACHES TO ADDRESSING GLOBAL ENVIRONMENTAL POLLUTION

The 20 strategies for reducing pollution in the global environment are listed below:

1. **Examining Pollution Systems, Processes, and Hotspots:** Collecting and analyzing data on pollutants in the environment is a key method for identifying pollution hotspots. It involves monitoring pollutant levels in soil, water, and air and tracking emissions from transportation sources and industrial processes. By carefully analyzing gathered data, researchers could uncover valuable insights into pollution levels, allowing them to identify trends, patterns, and potential pollution hotspots (Miettinen and Khan, 2022). One way to identify pollution hotspots is by using modeling tools and conducting on-the-ground assessments of a specific area's social and physical conditions. It involves analyzing the location of transportation routes and industrial facilities and evaluating the economic and health effects of pollution on nearby communities (Kanter et al., 2020).
2. **Proactive Measures to Minimize Waste Generation:** One way to achieve this is through extended producer responsibility initiatives, which hold manufacturers accountable for the environmental effects of their products. One effective approach is to minimize the consumption of disposable items and single-use plastics by promoting reusable packaging and containers. Additionally, implementing sustainable procurement practices, which prefer

products with little packaging made of recycled materials, can make a significant difference.

3. Improving the quality and yield of products through optimizing production processes involves a thorough analysis of each step in the production cycle. The overall productivity and output can be elevated by identifying areas where efficiencies can be enhanced and waste can be reduced. Manufacturers can boost their output by enhancing production flow, minimizing downtime, and optimizing processes without escalating their resource consumption. Ensuring the highest possible yield and quality of products is of utmost importance. It can be achieved by implementing effective QC measures, including setting product quality standards and conducting regular testing and inspection to ensure these standards are met (Ansori and Yusuf, 2023).
4. Enhancing the value of products through innovative product development, incorporating new technologies and features to enhance functionality and performance, and introducing new products that align with evolving emission standards and customer demands.
5. Engaging with stakeholders to incorporate their concerns and gather valuable feedback and input on environmental policies and initiatives. It includes gathering input on suggested environmental projects and understanding stakeholder viewpoints and concerns through focus groups, surveys, or other types of consultation (Liu et al., 2015).
6. Considering the economic concerns of stakeholders is crucial when implementing policies. It involves conducting a comprehensive economic impact assessment to analyze the potential risks, benefits, costs, and opportunities associated with an initiative for different stakeholders. This information is valuable for influencing decision-making and designing policies prioritizing economic benefits while minimizing negative environmental impacts.
7. Implement strategies to tackle environmental concerns and issues and roll out action programs to make a positive impact. These programs usually consist of a range of initiatives, policies, and measures designed to minimize the adverse effects of human activity on the environment and encourage sustainable practices. When implementing environmental action programs, it is important to consider the following steps: Identifying environmental issues, carrying out analysis and investigation to comprehend the problem's

origin, extent, and possible remedies, creating a plan of action, enlisting the help of various groups to engage stakeholders, carrying out the plan, and then monitoring and assessing its progress (Jin et al., 2022).

8. International environmental cooperation involves coordinating and collaborating with international organizations, countries, and other stakeholders to tackle global environmental issues. Global environmental challenges require international cooperation to address ocean pollution, biodiversity loss, and climate change effectively. These problems transcend national boundaries and cannot be solved by individual countries. Effective programs and policies can be developed by sharing best practices, resources, and knowledge.

9. Ensuring accountability and transparency in environmental decision-making requires public accessibility of environmental information. It also promotes public participation in ecological governance by empowering citizens to make well-informed decisions and actively contribute to shaping environmental policies (Amjad et al., 2022).

10. Enhancing the efficiency of current controls by maximizing the efficacy and value of regulations and measures that safeguard the environment. It includes strategies that enhance the implementation of current regulations, simplify enforcement processes, and strengthen reporting and monitoring mechanisms. Enhancing existing environmental controls is possible by adopting performance- and risk-based approaches. We can effectively address environmental concerns by focusing enforcement efforts on activities with the greatest environmental risks and establishing performance standards in particular sectors. Additionally, providing compliance assistance through training programs and technical support can help businesses meet environmental regulations. Regulatory reform includes updating and revising current regulations to ensure they are efficient, effective, and current by eliminating unnecessary or out-of-date requirements, expediting approval procedures, and enhancing the transparency and clarity of rules (Sauvé et al., 2016).

11. Implementing pollution control technologies is crucial for enhancing the effectiveness and efficiency of industrial processes, minimizing emissions and waste, and safeguarding human health and the environment. Various technologies can be utilized to address different types of pollution. For instance, air pollution control technologies like scrubbers, electrostatic precipitators, and particulate control systems can effectively remove pollutants from industrial exhaust gases. Similarly, water pollution control technologies such as filtration systems, sedimentation tanks, and wastewater treatment systems can eliminate pollutants from industrial wastewater before it is released into the environment. Solid waste management technologies like composting and recycling systems, leachate treatment systems, and landfill liners are crucial in minimizing waste generation and managing waste disposal. In addition, it is crucial to ensure that hazardous waste management technologies, including solidification and stabilization processes, chemical treatment systems, and incinerators, are designed to prioritize the disposal and safe management of hazardous waste (Sigmund et al., 2023).

12. Implementing an efficient taxation system for municipal waste disposal and ensuring effective landfill location and management are crucial aspects of environmental policy. Implementing fees or taxes on municipal waste disposal can provide economic incentives for businesses and individuals to minimize waste generation and promote reuse and recycling. This approach contributes to the conservation of natural resources and the reduction of GHG emissions. Implementing proper landfill management and siting techniques involves using engineered caps and liners, ensuring proper waste placement, and effective compaction methods. When siting landfills, it is crucial to consider factors like proximity to sensitive areas, hydrology, and geology.

13. Exploring the adoption of cutting-edge technologies to replace current pollution control methods, aiming to minimize environmental impact and enhance industrial processes. It is important to prioritize the development of technologies that can replace outdated pollution control techniques. These older methods may not be as effective or efficient in reducing pollution (Escher and Fenner, 2011).

14. Strict observance and implementation of the Stockholm Declaration's tenets across all nationalities. It includes safeguarding wildlife and natural resources, eliminating WMDs, avoiding the exhaustion of non-renewable resources, assisting developing countries in tackling pollution, preventing harmful pollution in oceans, organizing human settlements to avoid environmental problems, applying science and technology to enhance the environment, promoting essential environmental education, implementing suitable policies by governments, supporting environmental research in developing nations, fostering global collaboration on environmental issues, ensuring safe resource exploitation, and establishing national standards.

15. Changing to a Circular Economy: Reducing waste and increasing resource efficiency are the main objectives of a circular economy model. One important aspect of this work is to decrease the use of natural resources, encourage reuse and recycling, and create products that are durable and can be recycled. Implementing these measures can effectively decrease pollution linked to disposal, production, and extraction (Reed, 2008).
16. Advocating for the importance of environmental education and awareness through incorporating environmental education in school curricula, implementing awareness campaigns, and cultivating a sense of stewardship and responsibility within communities.
17. Conserving and restoring ecosystems through rehabilitating degraded areas and establishing protected areas to safeguard vital habitats is also vital.
18. Emphasizing the importance of urban planning and green infrastructure by creating sustainable transportation systems, green roofs, urban forests, and green spaces. These initiatives can enhance air and water and air quality and reduce pollution (Cho and Cheng, 2007).
19. Supporting innovation and research is crucial in finding solutions to global environmental challenges. It can be accomplished by backing scientific research, fostering technological advancements, and promoting interdisciplinary collaborations to create more efficient pollution control methods, alternative energy sources, and cleaner technologies.
20. Encouraging companies to use sustainable practices, minimize pollution in their operations, and foster accountability and transparency can substantially impact the pollution of the global environment by promoting social and corporate responsibility.

6.3. FUTURE PERSPECTIVES

The application of technology to solve environmental issues, such as improvements in water treatment technologies, is one possible avenue for policy development in the environmental sector. Various advancements in carbon capture and storage, energy storage, and renewable energy have the potential to decrease GHG emissions significantly. Given the growing awareness of the need to tackle environmental pollution, policy frameworks are expected to undergo significant changes. These changes will likely incorporate the latest research findings and foster global collaborations. With cutting-edge technologies like data analytics, remote sensing, and artificial intelligence, the efficiency and accuracy of pollution control efforts are expected to be significantly improved (Mashima and Okuyama, 2015).

6.4. SUMMARY

The pressing issue of global environmental pollution requires collaborative and urgent action. This chapter emphasizes the diverse range of pollution types, including noise, soil, water, and air pollution, all contributing to public health crises and environmental degradation. The main human causes of pollution include transportation, industrial activity, and agricultural practices. As a result, it is imperative to improve waste management techniques, enact sustainable farming methods, and implement cleaner technology. Implementing stringent EP and PEM is crucial in effectively addressing pollution. This chapter emphasizes the achievements of China's green growth, which can be attributed to strict enforcement of environmental policies. This success demonstrates the significant advancements that can be made in environmental policy and technology. The review acknowledges the difficulties in implementing policies, including economic barriers, non-compliance, and a lack of international cooperation. Despite the challenges, this chapter proposes forward-thinking solutions such as using satellite imaging for tracking pollution, fostering international research partnerships, and developing global environmental indices to improve pollution management. It suggests a comprehensive approach that supports SDGs, encouraging nations to enforce and implement comprehensive environmental policies and measures. Given the intricate nature of international leadership in environmental issues, it is crucial to emphasize the importance of strong institutions and carefully crafted policies that balance national sovereignty with global effectiveness. Through collective actions, we can make a substantial impact on reducing pollution and creating a cleaner environment for both present and future generations.

REFERENCES

1. Almer, C., & Winkler, R. (2017). Analyzing the effectiveness of international environmental policies: The case of the Kyoto Protocol. *Journal of Environmental Economics and Management, 82*, 125–151.
2. Amjad, S. O. H. A. I. L., Khan, A. S. I. F., & Usman, M. U. H. A. M. M. A. D. (2022). Analyzing international environmental law in addressing air pollution. *Asian Social Studies and Applied Research (ASSAR), 3*(01), 272–283.
3. Ansori, S., & Yusuf, R. (2023). Addressing the global environmental crisis: Strategies for sustainable development. *West Science Social and Humanities Studies, 1*(02), 63–75.
4. Appolloni, A., Sun, H., Jia, F., & Li, X. (2014). Green procurement in the private sector: A state-of-the-art review between 1996 and 2013. *Journal of Cleaner Production, 85*, 122–133.
5. Azizullah, A., Khattak, M. N. K., Richter, P., & Häder, D. P. (2011). Water pollution in Pakistan and its impact on public health—A review. *Environment International, 37*(2), 479–497.
6. Azzone, G., Noci, G., Manzini, R., Welford, R., & Young, C. W. (1996). Defining environmental performance indicators: An integrated framework. *Business Strategy and the Environment, 5*(2), 69–80.
7. Bernstein, J. A., Alexis, N., Barnes, C., Bernstein, I. L., Nel, A., Peden, D., & Williams, P. B. (2004). Health effects of air pollution. *Journal of Allergy and Clinical Immunology, 114*(5), 1116–1123.
8. Busscher, T., Zuidema, C., Tillema, T., & Arts, J. (2014). Bridging gaps: Governing conflicts between transport and environmental policies. *Environment and Planning A, 46*(3), 666–681.
9. Carlson, K. M., Gerber, J. S., Mueller, N. D., Herrero, M., MacDonald, G. K., Brauman, K. A., & West, P. C. (2017). Greenhouse gas emissions intensity of global croplands. *Nature Climate Change, 7*(1), 63–68.
10. Chamanee, G., Sewwandi, M., Wijesekara, H., & Vithanage, M. (2023). Global perspective on microplastics in landfill leachate: Occurrence, abundance, characteristics, and environmental impact. *Waste Management, 171*, 10–25.
11. Chambiwa, E., Majoni, J., & Mapuranga, L. L. (2022). The African Union and its role in ensuring African economic and environmental sovereignty. *Sovereignty Becoming Pulvereignty: Unpacking the Dark Side of Slave, 4*, 299–312.

12. Cho, W. C., & Cheng, C. H. (2007). Oncoproteomics: Current trends and future perspectives. *Expert Review of Proteomics, 4*(3), 401–410.
13. Clarkson, P. M., Li, Y., Richardson, G. D., & Vasvari, F. P. (2008). Revisiting the relation between environmental performance and environmental disclosure: An empirical analysis. *Accounting, Organizations and Society, 33*(4, 5), 303–327.
14. Dedić, A., Gerhardt, A., Kelly, M. G., Stanić-Koštroman, S., Šiljeg, M., Kalamujić Stroil, B., & Herceg, N. (2020). Innovative methods and approaches for WFD: Ideas to fill knowledge gaps in science and policy. *Water Solutions, 3*, 30–42.
15. DeJong, W., & Langford, L. M. (2002). A typology for campus-based alcohol prevention: Moving toward environmental management strategies. *Journal of Studies on Alcohol, Supplement,* (14), 140–147.
16. Di Vaio, A., Varriale, L., & Alvino, F. (2018). Key performance indicators for developing environmentally sustainable and energy efficient ports: Evidence from Italy. *Energy Policy, 122*, 229–240.
17. Dong, H., Yang, G., Zhang, Y., Yang, Y., Wang, D., & Zhou, C. (2022). Recycling, disposal, or biodegradable alternative of polyethylene plastic film for agricultural mulching? A life cycle analysis of their environmental impacts. *Journal of Cleaner Production, 380*, 134950.
18. Dwivedi, A. K. (2017). Researches in water pollution: A review. *International Research Journal of Natural and Applied Sciences, 4*(1), 118–142.
19. Elliott, L. (2000). ASEAN's environmental regime: Pursuing sustainability in Southeast Asia. *Global Environmental Change, 10*(3), 237–240.
20. Escher, B. I., & Fenner, K. (2011). Recent advances in environmental risk assessment of transformation products. *Environmental Science & Technology, 45*(9), 3835–3847.
21. Fell, H., & Maniloff, P. (2018). Leakage in regional environmental policy: The case of the regional greenhouse gas initiative. *Journal of Environmental Economics and Management, 87*, 1–23.
22. Fernando, C. S., Sharfman, M. P., & Uysal, V. B. (2017). Corporate environmental policy and shareholder value: Following the smart money. *Journal of Financial and Quantitative Analysis, 52*(5), 2023–2051.
23. Friedlingstein, P., O'Sullivan, M., Jones, M. W., Andrew, R. M., Gregor, L., Hauck, J., & Zheng, B. (2022). Global carbon budget 2022. *Earth System Science Data, 14*(11), 4811–4900.

24. Gasbarro, F., Rizzi, F., & Frey, M. (2013). The mutual influence of environmental management systems and the EU ETS: Findings for the Italian pulp and paper industry. *European Management Journal, 31*(1), 16–26.
25. Glencross, D. A., Ho, T. R., Camina, N., Hawrylowicz, C. M., & Pfeffer, P. E. (2020). Air pollution and its effects on the immune system. *Free Radical Biology and Medicine, 151*, 56–68.
26. Gupta, M. C. (1995). Environmental management and its impact on the operations function. *International Journal of Operations & Production Management, 15*(8), 34–51.
27. Hanks, E. H., & Hanks, J. L. (1969). An environmental bill of rights: The citizen suit and the National Environmental Policy Act of 1969. *Rutgers Law Review, 24*, 230.
28. Henri, J. F., & Journeault, M. (2008). Environmental performance indicators: An empirical study of Canadian manufacturing firms. *Journal of Environmental Management, 87*(1), 165–176.
29. Hermann, A., Chaudhuri, T., & Spagnol, P. (2005). Bipolar plates for PEM fuel cells: A review. *International Journal of Hydrogen Energy, 30*(12), 1297–1302.
30. Hertin, J., Berkhout, F., Wagner, M., & Tyteca, D. (2008). Are EMS environmentally effective? The link between environmental management systems and environmental performance in European companies. *Journal of Environmental Planning and Management, 51*(2), 259–283.
31. Hiremath, A. M., Tilwankar, A. K., & Asolekar, S. R. (2015). Significant steps in ship recycling vis-a-vis wastes generated in a cluster of yards in Alang: A case study. *Journal of Cleaner Production, 87*, 520–532.
32. Iwata, H., & Okada, K. (2014). Greenhouse gas emissions and the role of the Kyoto Protocol. *Environmental Economics and Policy Studies, 16*, 325–342.
33. Jin, L., Apte, J. S., Miller, S. L., Tao, S., Wang, S., Jiang, G., & Li, X. (2022). Global endeavors to address the health effects of urban air pollution. *Environmental Science & Technology, 56*(11), 6793–6798.
34. Johnson, J. M. F., Franzluebbers, A. J., Weyers, S. L., & Reicosky, D. C. (2007). Agricultural opportunities to mitigate greenhouse gas emissions. *Environmental Pollution, 150*(1), 107–124.
35. Kanter, D. R., Chodos, O., Nordland, O., Rutigliano, M., & Winiwarter, W. (2020). Gaps and opportunities in nitrogen pollution policies around the world. *Nature Sustainability, 3*(11), 956–963.

36. Karagozoglu, N., & Lindell, M. (2000). Environmental management: Testing the win-win model. *Journal of Environmental Planning and Management, 43*(6), 817–829.

37. Khan, M. Z. H. (2017). Environmental management system of textile industry in Bangladesh: Constraint and remediation. *International Journal of Environment, Ecology, Family and Urban Studies, 7*(2), 1–12.

38. Lawrence, D. P. (2000). Planning theories and environmental impact assessment. *Environmental Impact Assessment Review, 20*(6), 607–625.

39. Le Quéré, C., Andres, R. J., Boden, T., Conway, T., Houghton, R. A., House, J. I., & Zeng, N. (2013). The global carbon budget 1959–2011. *Earth System Science Data, 5*(1), 165–185.

40. Liefferink, D., & Andersen, M. S. (1998). Strategies of the 'green' member states in EU environmental policy-making. *Journal of European Public Policy, 5*(2), 254–270.

41. Liu, J., Mooney, H., Hull, V., Davis, S. J., Gaskell, J., Hertel, T., & Li, S. (2015). Systems integration for global sustainability. *Science, 347*(6225), 1258832.

42. Mashima, R., & Okuyama, T. (2015). The role of lipoxygenases in pathophysiology: New insights and future perspectives. *Redox Biology, 6*, 297–310.

43. McCuin, R. M., & Clancy, J. L. (2003). Modifications to United States Environmental Protection Agency methods 1622 and 1623 for detection of Cryptosporidium oocysts and Giardia cysts in water. *Applied and Environmental Microbiology, 69*(1), 267–274.

44. Melnyk, S. A., Sroufe, R. P., & Calantone, R. (2003). Assessing the impact of environmental management systems on corporate and environmental performance. *Journal of Operations Management, 21*(3), 329–351.

45. Meybeck, M., & Helmer, R. (1989). The quality of rivers: From pristine stage to global pollution. *Palaeogeography, Palaeoclimatology, Palaeoecology, 75*(4), 283–309.

46. Michelsen, O., & de Boer, L. (2009). Green procurement in Norway: A survey of practices at the municipal and county level. *Journal of Environmental Management, 91*(1), 160–167.

47. Miettinen, M., & Khan, S. A. (2022). Pharmaceutical pollution: A weakly regulated global environmental risk. *Review of European, Comparative & International Environmental Law, 31*(1), 75–88.

48. Panwar, R., Hansen, E., & Kozak, R. (2014). Evaluating social and environmental issues by integrating the legitimacy gap with expectational

gaps: An empirical assessment of the forest industry. *Business & Society, 53*(6), 853–875.

49. Ponce, P., Oliveira, C., Álvarez, V., & del Río-Rama, M. D. L. C. (2020). The liberalization of the internal energy market in the European Union: Evidence of its influence on reducing environmental pollution. *Energies, 13*(22), 6116.
50. Reed, M. S. (2008). Stakeholder participation for environmental management: A literature review. *Biological Conservation, 141*(10), 2417–2431.
51. Roy, P., Nei, D., Orikasa, T., Xu, Q., Okadome, H., Nakamura, N., & Shiina, T. (2009). A review of life cycle assessment (LCA) on some food products. *Journal of Food Engineering, 90*(1), 1–10.
52. Sarkar, R. (2008). Public policy and corporate environmental behaviour: A broader view. *Corporate Social Responsibility and Environmental Management, 15*(5), 281–297.
53. Sauvé, S., Bernard, S., & Sloan, P. (2016). Environmental sciences, sustainable development, and circular economy: Alternative concepts for trans-disciplinary research. *Environmental Development, 17*, 48–56.
54. Schleussner, C. F., Rogelj, J., Schaeffer, M., Lissner, T., Licker, R., Fischer, E. M., & Hare, W. (2016). Science and policy characteristics of the Paris Agreement temperature goal. *Nature Climate Change, 6*(9), 827–835.
55. Schwarzenbach, R. P., Egli, T., Hofstetter, T. B., Von Gunten, U., & Wehrli, B. (2010). Global water pollution and human health. *Annual Review of Environment and Resources, 35*(1), 109–136.
56. Scott, S. V. (2015). Does the UNFCCC fulfil the functions required of a framework convention? Why abandoning the United Nations Framework Convention on Climate Change might constitute a long overdue step forward. *Journal of Environmental Law, 27*(1), 69–89.
57. Sigmund, G., Ågerstrand, M., Antonelli, A., Backhaus, T., Brodin, T., Diamond, M. L., & Groh, K. J. (2023). Addressing chemical pollution in biodiversity research. *Global Change Biology, 29*(12), 3240–3255.
58. Steinebach, Y., & Knill, C. (2017). Still an entrepreneur? The changing role of the European Commission in EU environmental policy-making. *Journal of European Public Policy, 24*(3), 429–446.
59. Tatoglu, E., Bayraktar, E., & Arda, O. A. (2015). Adoption of corporate environmental policies in Turkey. *Journal of Cleaner Production, 91*, 313–326.

60. Van der Oost, R., Beyer, J., & Vermeulen, N. P. (2003). Fish bioaccumulation and biomarkers in environmental risk assessment: A review. *Environmental Toxicology and Pharmacology, 13*(2), 57–149.
61. Zeithaml, C. P., & Zeithaml, V. A. (1984). Environmental management: Revising the marketing perspective. *Journal of Marketing, 48*(2), 46–53.
62. Zhang, K. M., & Wen, Z. G. (2008). Review and challenges of policies of environmental protection and sustainable development in China. *Journal of Environmental Management, 88*(4), 1249–1261.

Chapter 7

EQUITY, ENVIRONMENTAL JUSTICE, AND URBAN CLIMATE CHANGE

Contents

INTRODUCTION

This chapter focuses on equity and environmental justice aspects of climate change in and across cities. Within cities, climate change equity and environmental justice aspects are important due to distinctive urban characteristics related to socioeconomic and ethnic diversity as well as to high density (UN-Habitat, 2010). Stratification is common in diverse societies, where individuals are categorized based on various factors such as ability, age, ethnicity, race, profession, gender, caste, and social class. These social categories within urban populations significantly impact the ability of groups and individuals to withstand climate stresses and reduce or adapt to climate risks. There is a concern that differences between strata can result in unfair treatment and discrimination of individuals based on their group membership, which raises equity issues. For instance, individuals with lower incomes and from ethnic minority backgrounds often reside in areas that are densely populated and more susceptible to the impacts of climate change, which reduces their ability to cope with risks (Ikeme, 2003).

Environmental justice and Climate change equity issues become apparent when considering cities that historically emitted low levels of GHGs per capita yet now face above-average impacts of climate change. This is particularly true for many cities in low-income countries. For instance, many mega-disasters took place from January 1975 to October 2008. These disasters, which resulted in a high number of fatalities, primarily impacted countries with low-income levels. Surprisingly, despite accounting for only a small percentage of the total number (0.26%) of events, these disasters caused a staggering 78% of the total mortality rate (Amorim et al., 2022). Urban areas experiencing significant climate change impacts are witnessing rapid population growth, particularly in low-income countries of Africa and Asia. It is projected that nearly 90% of the urban population increase by 2050 will occur in these regions.

These trends pose a growing threat to many individuals as climate change escalates. It could potentially exacerbate environmental justice and equity issues, particularly due to the rapid urbanization in Africa and Asia, which is expected to lead to a rise in slum populations. It is worth noting that approximately 62% of the population in Africa and approximately 30% in Asia reside in slums, according to Schlosberg and Collins (2014). Rapid urban growth presents a significant problem for city managers, even though the growth of informal densely populated settlements, frequently driven by rural-urban migration, indicates expected development and, hence, a city's "success." It is frequently linked to unregulated and unplanned settlements in risk-prone areas.

Climate change exacerbates equity issues in high-income countries' cities, as demonstrated by the devastating heat waves in Central Europe in 2003, resulting in a significant number of deaths among the elderly population. A similar pattern was observed in Chicago in 1992, where predominantly lower-income neighborhoods experienced a high number of excess deaths. Melbourne also faced a tragic toll in 2009, with a considerable number of heat-related fatalities (Wilson et al., 2010).

7.1. DEFINITIONS, PRINCIPLES, AND DOMAINS

Many regional and local climate initiatives aim to promote equity, which is sometimes stated explicitly. But it's not always clear to what extent equity is being discussed: between households within communities, between national stakeholders, local groups, and urban districts, or between generations of urban residents; between individuals, like men and women, or in the distribution of burdens and benefits. These aspects pertain to a specific dimension of equity concerns, specifically focusing on outcomes (Wiersum et al., 2014). Meanwhile, scholars have identified three distinct dimensions related to this topic:

i. Consequential, distributive, or Outcome-based equity.

ii. Procedural or process-oriented equity has been discussed by various researchers.

iii. Contextual equity.

Considering the impacts of a developmental, action, or policy trend on both high- and low-income countries is crucial when discussing outcome-based equity. Procedural equity is all about ensuring impartiality and fairness in administering and delivering justice. This topic is frequently addressed in the context of urban climate programs in low-income countries. Considering the pre-existing social, economic, and political conditions, contextual equity links the first two dimensions together.

McDermott et al. (2011) provide a framework for operationalizing equity by linking its dimensions to three key variables: process, goal, and target (Figure 7.1). Operationalization is also guided by various indicators and principles, with numerous proposals (Table 7.1).

There are variations in the level of support for operational indicators and equity principles among different countries and potentially within local and subnational levels. The "polluter pays" theory was most supported by delegates to the UNFCCC climate change negotiations, at least from a short-term (i.e., ≤20 years) perspective. Next came the consideration of "ability to pay" and "exemption of the poorest." Many delegates did not support the idea of equal mitigation pledges and even more objected to the concept of sovereignty, which refers to a state's right to govern itself and make decisions about its mitigation pledges. According to Cazorla and Toman (2000), there is unlikely a simple solution to equity disputes in the context of international climate policies. Resolving the question of which policies will be equitable in the long term will require significant effort and time. The Nationally Determined Contributions (NDCs) provide an alternative solution to the issue of who should pay and how much, following the Paris Agreement and COP21. Now, every nation is showcasing its voluntary pledges to reduce GHG emissions, fostering a sense of ownership and facilitating compliance enforcement (Janin and Chothia, 1985).

Table 7.1. Frequently Used Equity Domains in Global Efforts to Mitigate Climate Change

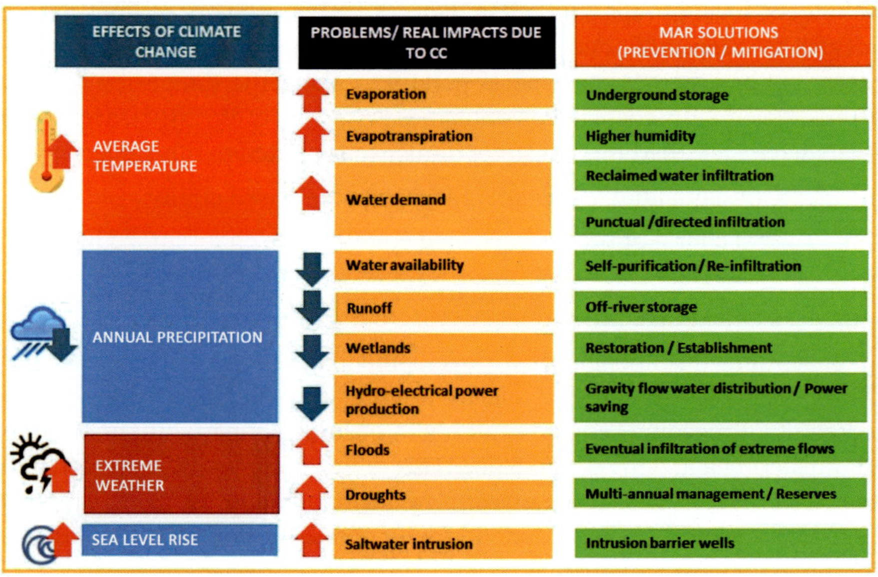

Source: Escalante EF, Sauto JSS, Gil RC. (2019). Sites and indicators of MAR as a successful tool to mitigate climate change effects in Spain. Water, 11(9):1943. https://doi.org/10.3390/w11091943.

7.1.1. Equity and Climate Change

Climate change discussions began incorporating concerns of equality, equity, and environmental justice when it became evident that countries with the least historical contribution to global warming could face the greatest future impacts. However, it is important to note that due to interdependencies, high-income countries are also likely to experience increased impacts. Therefore, the early conversations focused on the responsibility for mitigating environmental issues and the commitments made by various parties. According to Thomas and Twyman (2005) it is important to focus on mitigation when discussing equity concerning climate change and consider the adaptation and impacts. Understanding the significance of this issue is crucial for urban areas, as it is at the regional and local levels where we will witness varying adaptation and impact requirements. The topic of justice in adaptation has been receiving increased attention in recent discussions. Furthermore, all three aspects are relevant because mitigation and adaptation lessen climate change's effects and risks (*see* Figure 7.2).

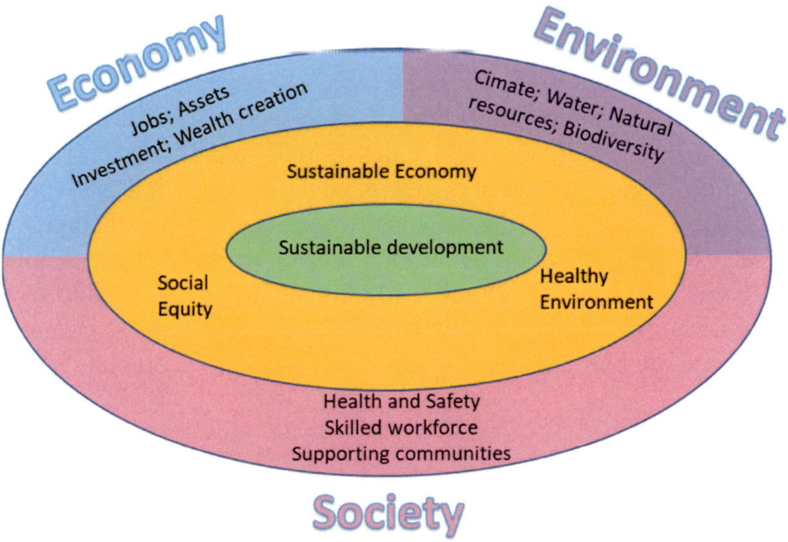

Figure 7.1. *Thematic representation of the equity framework (Source: Salvador Costa MJ, Leitão A, Silva R, Monteiro V, Melo P. (2022). Climate change prevention through community actions and empowerment: A scoping review. International Journal of Environmental Research and Public Health.; 19(22):14645. https://doi.org/10.3390/ijerph192214645, Creative Commons License.*

Climate change affects individuals and communities in various ways, influenced by socioeconomic conditions, including assets, income, and discrimination based on impaired mobility, poor health, age, gender and sex, ethnicity or race, and minority status. The characteristics mentioned significantly impact people's residences and the extent to which they are impacted. Nevertheless, the impact is not limited to only the most vulnerable individuals. Regular events can also slowly erode the resource base of stronger communities in society, ultimately resulting in greater levels of urban poverty (McManus et al., 2014).

It is often observed that areas with high sensitivity and exposure to climate impacts tend to have limited adaptive capacity, as highlighted by The World

Bank in 2010. It is crucial to acknowledge the present vulnerability of numerous cities in middle- and low-income nations and the inability of their governments and residents to adjust to a shifting climate. Deficiencies in location, housing quality, infrastructure, facilities available, social services, education access, and opportunities, the effectiveness of planning systems, as well as low levels of community, limited resources, and individual adaptive capacity, often contribute to differential vulnerability. "Risk accumulation processes" and repeated coping also gradually reduce adaptive capacity, which has repercussions for chronic poverty (Smit and Pilifosova, 2003).

Contemporary urban planning is increasingly concerned about mitigation issues due to the significant contribution of urban areas to global CO_2 emissions. According to measurement protocols, cities of high- and middle-income nations are responsible for approximately 70% of the total emissions. Several cities have demonstrated visionary leadership by setting targets and implementing plans to reduce greenhouse gas emissions. These efforts have been supported by city networks like the ICLEI and C40 Cities Climate Leadership Group — Local Governments for Sustainability. However, assessing these commitments in terms of how the burdens and benefits are distributed is crucial (Schipper et al., 2021).

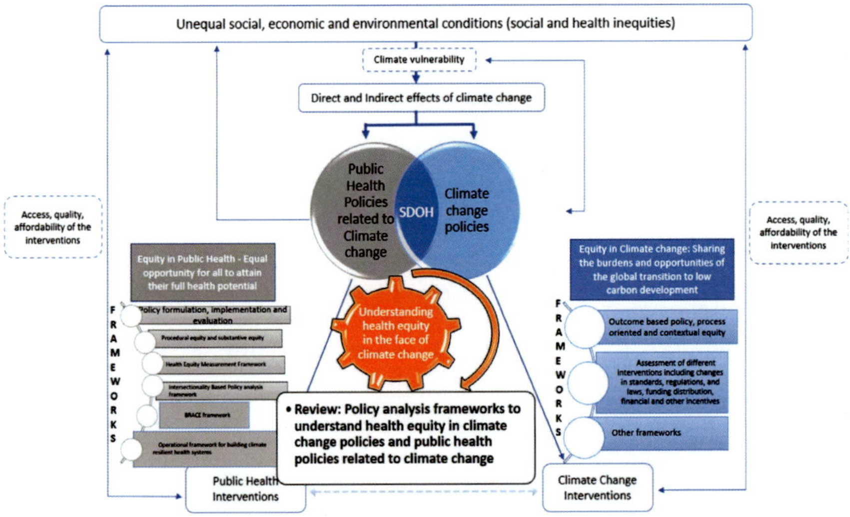

Figure 7.2. Conceptualization of equity in climate policy (Source: Khanal S, Ramadani L, Boeckmann M. (2022). Health equity in climate change policies and public health policies related to climate change: Protocol for a systematic review. International Journal of Environmental Research and Public Health, 19(15):9126. https://doi.org/10.3390/ijerph19159126, Creative Commons License.

7.2. EQUITY, URBAN IMPACTS, AND ADAPTATION

Extreme weather events and climate change can have varying impacts on urban residents, depending on various factors. Location plays a significant role in determining the physical exposure of a community, while inherent or "constructed risk" and urban development processes and inherent or "constructed risk" are also important factors to consider. There is a connection between risk and the demographic, economic, and social characteristics of a population. Institutions, governance, and power dynamics influence this vulnerability. Many of these factors are interconnected. Due to modern and historical urban development processes, they take place in large, medium, and small cities as well as in high-, middle-, and low-income countries (Chu et al., 2017).

7.2.1. Exposure, Social Context, Demographics, Governance, and Power

It is becoming increasingly clear that the effects of climate change, both extreme and gradual, have a greater impact on individuals with lower social status and incomes. This effect is particularly evident in cities in South America and India, as highlighted by various studies. It is worth mentioning that the impact of environmental issues extends beyond just the most vulnerable individuals. Even those better off in society are affected by floods and droughts, which can undermine their resource base. Certain individuals, such as infants, young children, young infants, and older age groups with impaired mobility, may be more vulnerable to multiple risks (Adams et al., 2011).

Locational and geographic factors play a significant role in shaping climate risks. In many cities in middle- and low-income nations, residents with low incomes and social status tend to live in areas more vulnerable to climate risks, such as flood plains and low-elevation coastal zones. It is evident in Dhaka, Rio de Janeiro, Alexandria, Cairo, Lagos and other Asian cities. Illegal housing construction is common in high-risk urban areas, where building codes are often disregarded. These areas often have dense populations and substandard buildings, which puts residents from lower social classes at risk of climate-related hazards and potentially devastating consequences (Mabillard, 2022).

As a component of the City Prosperity Index (CPI), Bennett et al. (2016) measures equity and social inclusion, which is represented in Figure 7.3.

1. The indicators and dimensions considered include quality of life, environmental sustainability, social inclusion, and economic equity. Economic equity is measured by factors such as the poverty rate and urban Gini coefficient. Social inclusion is assessed through indicators

like slum households and youth unemployment. Gender inclusion is evaluated based on equitable secondary school enrollment.

2. Environmental sustainability is measured by air quality, waste management, and energy usage.
3. Quality of life is assessed by considering factors like health, education, and safety and security. The other three dimensions of the CPI, namely Governance and Legislation, Infrastructure, and Productivity, are not displayed due to their lack of relevance or absence of data (May, 2019).

Looking at the case studies in Figure 7.3, it is evident that equity and social inclusion tend to be lacking in cities of Southern and Central Africa and in cities of South America. Within African cities, there is a concerning lack of equity and social inclusion, which unfortunately leads to a relatively low quality of life. It manifests in various ways, including inadequate access to healthcare, education, and safety measures. Climate change impacts are likely to exacerbate this situation, although the evidence and confidence are limited. Dar es Salaam stands out from the typical African pattern, with a relatively high level of equity but a low quality of life. On the other hand, South American cities maintain a moderate quality of life despite disparities, although the impact of climate risks on these cities remains uncertain (Feezell, 2018).

There are variations in equity and social inclusion levels across different regions. North American cities tend to have higher levels than South American cities, but they still fall below the average levels observed in European cities. Moscow stands out in Europe as a city with low equity and social inclusion. These criteria in cities across Asia are quite varied. In high-income nations, cities like Melbourne and Tokyo strongly focus on equity. Equity levels in middle-income cities like Dhaka, Jakarta, and Manila tend to be somewhat lower. In these cities, the equity index surpasses the index for quality of life. A significant portion of the urban population is residing in substandard living conditions (Liu and Ma, 2016).

Recognizing that these interpretations are based on a relatively small sample, caution should be practiced against concluding the case study cities as a whole (based on data from UN-Habitat State of the World Cities 2012/2013 – Prosperity of Cities). Investment, governance, and legal practices play a significant role in shaping urban risks. It is widely recognized that low-income urban populations are particularly vulnerable to the devastating effects of extreme weather events. This vulnerability is primarily attributed to these communities' inadequate services and infrastructure, such as access to emergency and healthcare services, all-weather roads, solid waste collection, drainage, sanitation, and piped water (Kammerbauer and Wamsler, 2017). The impact on low-income urban neighborhoods is often exacerbated by a lack of capability in urban governments or their unwillingness to address significant service and infrastructure deficits. This issue is prevalent not only in middle- and low-income countries but also in high-income countries. Hurricane Sandy in New York and Hurricane Katrina in New Orleans had a greater impact on social groups with lower social status and incomes, specifically women and ethnic minorities. Nevertheless, the risks faced by numerous residents in most cities in middle- and low-income countries significantly impact a larger portion of the urban population (Ma and Christensen, 2019).

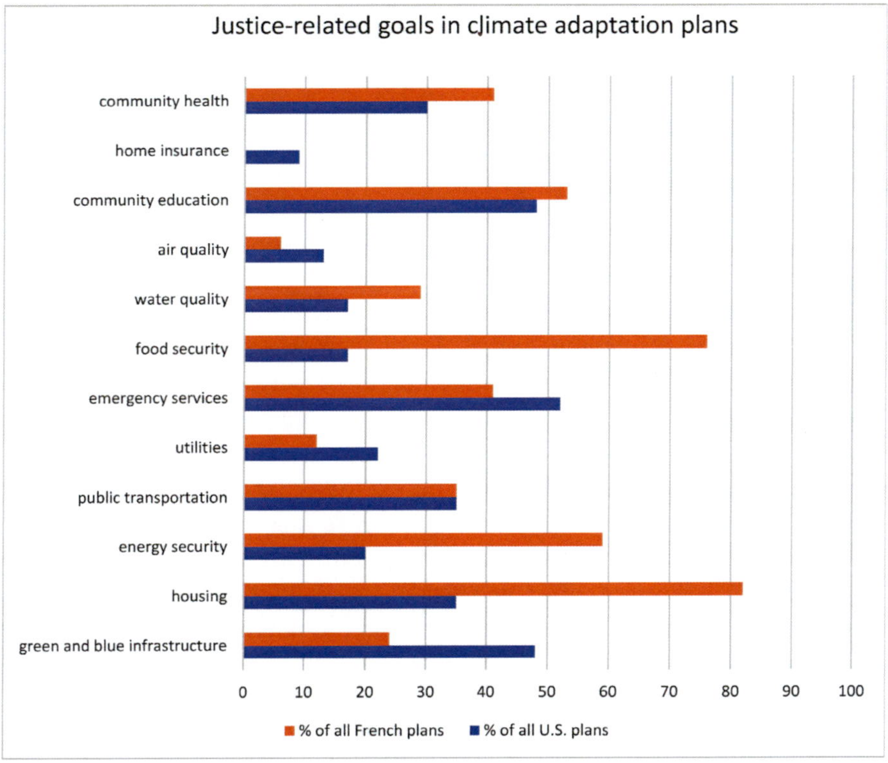

Figure 7.3. Justice-related goals in various domains of climate adaptation plans in the U.S. and French plans.

Source: Lioubimtseva E, da Cunha C. (2023). The role of non-climate data in equitable climate adaptation planning: lessons from small French and American cities. Sustainability.; 15(2), 1556. https://doi.org/10.3390/su15021556, Creative Commons License.

When the government fails to acknowledge and tackle current vulnerabilities, it creates social risk or what some refer to as "development-accumulated" risk.

Recent local initiatives have made commitments to include environmental justice and equity in their plans to enhance resilience to climate change. One example is the "OneNYC Plan: The Plan for a Strong and Just City" implemented by New York in 2015. Unfortunately, in many cities, middle- and low-income neighborhoods are still situated in areas prone to hazards due to past development patterns. The infrastructure in these neighborhoods is particularly vulnerable to climate shocks. It is common to find shortcomings in the infrastructure designed to reduce risks and respond to disasters. Unfortunately, the lessons learned from past disasters often receive insufficient attention. Over time, there has been a concerning accumulation of risk, as evidenced by the growing number of disaster losses in cities caused by various natural events such as and tropical storms, tsunamis, earthquakes, floods, mega-debris flows in the last 20 years (Ben Selma et al., 2022).

7.2.2. Equity and Climate Hazards

7.2.2.1. Heat

Heat-related impacts are a significant concern regarding climate change in urban areas. The rise in the average global temperature and the urban heat island (UHI) effect—the difference in temperature between densely populated places and their surrounding rural areas—converge. These dynamics can positively mitigate the risks associated with cold temperatures, but they can also create specific risk situations during periods of heat waves or extreme heat. Heat waves are extended periods of intense heat that surpass a specific temperature threshold above the average long-term temperature for a particular city. These thresholds are often measured in standard deviations from the mean temperature. Heat waves are a significant climate-related risk, with more fatalities from heat waves than other climate hazards like hurricanes and floods (Luber and McGeehin, 2008).

Heat waves in urban areas can lead to higher rates of illness and death due to the direct impact of heat stress and related factors. High night-time temperatures can harm the human body, preventing regeneration and rest, especially when it comes to direct heat stress. Heat and relative environmental factors, such as water and air pollution, can indirectly affect health. For instance, the presence of air pollutants and heat can lead to increased ozone levels, which can negatively affect lung function and cause irritation to the respiratory system. Heat waves have the potential to trigger heart attacks and worsen respiratory conditions such as bronchitis, asthma, and various cardiopulmonary diseases, ultimately resulting in untimely fatalities. The combined impact of direct heat stress and ozone is significant (Chen et al., 2016). Heat-related risks can potentially impact all citizens equally, as ozone and heat impact the general urban environment. Heat-related risk varies among different groups and is influenced by individual characteristics (intrinsic) and socioeconomic factors (extrinsic). Physiological characteristics, including sex, age, disability, and health condition, are examples of intrinsic factors. External factors include location-related, ecological, and societal attributes such as economic standing, gender, and the quality of living and working environments.

Based on a comprehensive analysis of multiple studies, it has been found that age plays a significant role in increasing the risk of heat-related mortality. Various studies have published relative risk ratios ranging from 1.3 to 3.7, highlighting the impact of age as an intrinsic factor. A single study indicates a higher relative risk ratio for children compared to the general population (Webb, 1995).

Compared to men, women are more likely to die from heat-related causes (RRR $\geq$ 1.0–1.4). Research has shown that there may be thermo-regulatory and physiological differences between women and men, which could make women more heat intolerant. On the other hand, women often face higher heat exposure levels than men. This is primarily because they spend more time in enclosed spaces, such as cooking in poorly ventilated houses without proper air-conditioning or air flow (Chapman and Pollack, 1975).

People who are less mobile and confined to bed have a higher heat wave vulnerability based on medical status. This vulnerability is indicated by an odds ratio (OR) of 3–9, as shown by studies by (Sun et al., 2019). Individuals with cardiovascular diseases face a significantly increased risk, with the overall population having a risk ratio of 4.05 and the elderly facing an even higher risk ratio of up to 34.1.

Location-specific characteristics are the main socioeconomic and extrinsic factors reported in the literature for heat-related effects. While there is a lack of systematic reporting on the correlation between socioeconomic levels and relative risk, the general pattern suggests that individuals with lower socioeconomic status and education levels are more vulnerable to heat stress. This vulnerability is determined by household ownership, overcrowding, household size, unemployment, occupation, and education data. According to Hasterok et al. (2011), the impact of heat is greater on socioeconomically disadvantaged households. This is due to their location in areas with limited access to urban green infrastructure and limited resources to maintain and create private green spaces. Open spaces and bodies of water, like pools, have been found to create a cooling effect in their surroundings, which helps to reduce risks.

When examining risks in different urban areas, it is evident that individuals residing in inner cities face higher risk levels than those who live in the suburbs. Climate variations impact the risks faced by different regional or urban areas. According to the research on the 2003 heat wave in Europe, it was found that death rates in London increased by 42%, which was significantly higher than the average increase of 16% observed in Wales and England as a whole. There is a potential correlation between excessive mortality in London and the urban heat island effect, which may amplify the impact of climatic heat stress (Stein and Stein, 1994).

Living and working conditions can significantly impact susceptibility to heat waves, particularly for individuals in low-income households who work from home and experience heat stress.

Dong et al. (2014) found an odds ratio of 1.86 in Ahmedabad, India, as (Macdonald, 1993) mentioned. Residents living under the roof and on the upper floor face significant risks, with a relative risk reduction of 5.4 and 4.7, respectively. In addition, research indicates that loneliness, especially among elderly individuals and unmarried men, and associated behaviors can be considered potential risk factors. Given the combined impact of various factors, it has been observed that indoor heat stress is more common among certain vulnerable groups, such as older people, children, and women who reside in substandard housing conditions (Pollack et al., 1993).

Efforts to support vulnerable populations at risk of heat-related mortality can take into account both extrinsic and intrinsic factors. Considering the importance of the matter, it is crucial to provide individuals at risk with wider social support and specialized healthcare. For instance, during periods of excessive heat, it would be beneficial for community group members or neighbors to check in on elderly individuals living alone. While air conditioning is widely used for managing heat, it is important to consider its sustainability (Forman et al., 2016). Many existing systems utilize electricity and contribute to greenhouse gas emissions if fossil fuels power them. In addition, the high running and purchase costs may prevent many low-income residents from being able to use it. In addition, extreme heat can lead to electricity brown-outs or black-outs, causing a complete lack of access to air conditioning. Public health managers and urban planners can enhance their support to disadvantaged households by implementing measures at the extrinsic level. It encompasses modern and traditional adaptive architecture in high-risk areas and heat-trapping technology that cools instead of trapping heat (Lindzen and Farrell, 1980) (e.g., through landscape design, greening, and albedo adjustments).

7.2.2.2. Rain

Various risks to human well-being are associated with precipitation-related hazards, including drought, landslides, and flooding. Inland flooding has the potential to cause widespread devastation, as seen in various instances such as in Thailand in 2011, Australia in 2011, and Pakistan in 2010 (Trenberth and Zhang, 2018). However, localized floods can have significant impacts, posing threats to livelihoods, lives, and health. Many cities have seen the emergence of informal settlements in areas prone to landslides or flooding due to their geographical characteristics. Mistimed rainfall and insufficient precipitation can have a significant impact on urban residents. These events can increase food prices, crop failures, and water shortages, which are particularly detrimental to low-income populations (Garg and Nayar, 2005).

7.2.2.3. Inland Flooding

Urban flood risk in middle- and low-income countries arises from various factors. These include impermeable surfaces that hinder water absorption, leading to rapid run-off. The lack of parks and green spaces also exacerbates the issue by limiting water absorption areas. Inadequate drainage systems, often obstructed by waste, quickly become overwhelmed with water. Furthermore, the irrational decision to develop housing on natural buffers, including marshlands, contributes to the problem (Messager et al., 2021).

Urban residents with lower incomes are frequently at a higher risk of being exposed to environmental hazards than other city dwellers. This is because affordable housing options are often situated in areas more prone to environmental risks, such as slopes and floodplains. Additionally, these neighborhoods often lack proper waste management services and drainage systems, as municipal governments overseeing them have failed to establish and maintain such infrastructure. 40% of Asian urban dwellers, according to the Asian Development Bank (ADB) (2010), live in slums or subpar housing, which are frequently located along a city's canals and rivers. Since these areas are usually owned by the government, settling there is usually less problematic than on private land (Thieken et al., 2022). Living near urban waterways often occurs due to the high demand for land in rapidly expanding cities. This situation is particularly prevalent among the urban new and poor migrants who lack secure tenure. Unfortunately, this can disrupt social support networks and livelihoods, as well as population displacement when flooding happens repeatedly. Additional consequences arise from unhygienic environments and substandard housing, as highlighted by (Pietrafesa et al., 2019). When flooding happens, there is a high chance of hazardous materials contaminating wells and open waters. This increases the risks of skin, respiratory, and water-borne diseases.

7.2.2.4. Landslides

Landslide risks have not been given the same level of attention as coastal hazards and flooding. According to Jongmans and Garambois (2007), landslides are often linked to other natural hazard triggers, including floods, earthquakes, or extreme precipitation in the databases of natural disasters. This contributes to diminishing the concern and awareness of the general public and authorities regarding the risk of landslides. However, landslides pose substantial risks to human welfare in numerous urban areas.

Rainfall-triggered landslides result from a combination of locational and geo-hydrological factors. The intensity and duration of precipitation are key factors in understanding geo-hydrological processes. Factors influencing location selection include vegetation cover, soil moisture, and susceptibility to fracturing, rock strength, and slope. These factors were identified in risk analyses conducted in El Salvador, with comparisons made to Sri Lanka, Nepal and Indonesia. These studies utilized DESINVENTAR3 data to analyze the frequency of rainfall-induced landslides and their correlation with socioeconomic and population density composition. Nonetheless, the research reached varied outcomes when the geo-hydrological exposure was maintained constant (Korup et al., 2010).

In El Salvador, municipalities with higher socioeconomic status had higher rates of landslide mortality compared to those with lower socioeconomic status. In contrast, the analysis conducted in Indonesia revealed a clear negative association between landslide mortality and the Human Development Index (HDI), even after accounting for physical exposure. Based on similar methods, these examples highlight the importance of being cautious when drawing generalizations about the relationship between vulnerability to landslides and poverty (Malamud et al., 2004).

It is important to exercise caution when implementing policy measures motivated by equity concerns. Public policies and infrastructure measures to upgrade or protect settlements may inadvertently exacerbate risks in these areas or nearby neighborhoods. For instance, in Medellín, there has been criticism regarding infrastructure improvements and their potential negative impact on environmentally protective infrastructure, like parks on hill slopes that play a crucial role in reducing the risk of flooding and landslides (Korup et al., 2010). Run-off to the low-lying districts of Rio de Janeiro has increased due to the paved walkways in favelas redevelopment of slums. Several studies conducted in Rio have examined the relationship between soil hydrology, rainfall intensity, and slope instability as factors influencing landslides. This remains true regardless of whether or not the slopes are disturbed. The distribution of landslide risk is uneven due to the combination of locational and geophysical factors, which are influenced by patterns of deprivation. Despite historical conflicts and the control of certain areas by organized gangs, inadequate infrastructure investments and a lack of public policies have resulted in significant equity issues in those cities and their poorer neighborhoods (Rodrıguez et al., 1999).

7.2.2.5. Drought

It is often overlooked that numerous cities in middle- and low-income countries are situated in dryland ecosystems characterized by unpredictable and low precipitation. According to a study by Wilhite and Glantz (1985), approximately 45% of the dryland population is classified as urban. According to Keyantash and Dracup (2002), dryland ecosystems span approximately 41% of the Earth's surface and support a population of around 2 billion. Middle- and low-income nations in the drylands make up around 72% of the total land and 87–93% of the population.

Urban areas can experience a range of consequences from drought, such as electricity and water shortages (in cases where hydropower is used), water-borne diseases from contaminated water, and higher food prices resulting from reduced supplies. Approximately 150 million individuals currently reside in cities facing ongoing water scarcity, where the sustainable groundwater and surface flow within their urban areas is below 100 liters per person per day. According to a study by Svoboda et al. (2002), the global population is projected to reach 1 billion by 2050.

It is worth noting that there may be a disproportionate representation of women among those who will face perennial water shortages in the future. This is because women often belong to the most economically disadvantaged groups. It is worth noting that the described impacts, such as food insecurity and a rise in food prices in urban areas (Chang and Kleopa, 1991), will have a greater impact on women. This is because they consume less food than other family members when food becomes scarce or expensive (Mukherjee et al., 2018).

Urban areas will probably face the dual challenges of water shortages and drought, which severe and destructive rainfall events will exacerbate. According to Kogan (1997), the annual rainfall is usually concentrated in a few erosive and intense storms in dryland ecosystems. In the Sahel, a severe flood occurred in 2010, which had a devastating impact on the middle Niger River valley. Niamey was particularly affected, with approximately 5,000 people losing their homes due to the inundation of 3.1 square kilometers. Intense rainfall, although not common in that area, was the cause of the damage. The main reason for the destructive impact was the region's land-use changes. In February 2015, São Paulo experienced severe flooding despite being in the middle of a serious drought. This incident serves as another example of the unpredictable nature of weather patterns. Within a single hour, São Paulo was hit with a staggering 96 millimeters of rainfall, coinciding with the most extreme drought the city has faced in the last eight decades. It's interesting to note that despite the heavy rainfall causing flooding and traffic problems in the city, unfortunately, it did not contribute sufficiently to replenish the reservoirs outside the urban area. As a result, water rationing needed to be maintained (Haile et al., 2020).

When droughts coincide with other extreme events, their consequences are amplified, leading to a compounding impact. This is particularly challenging for cities already grappling with the aftermath of a previous event. Intensive flooding caused by heavy rainfall in droughts is just one example. Droughts also raise the fire risks in heat-sensitive ecosystems surrounding water-stressed cities (Wilhite et al., 2014).

In addition, the impact of climate change on long-term drought risk can potentially affect water availability in cities, as seen in the case of Mexico City. There are concerns about equity in this situation, as it has been found that lower-income individuals spend a significantly higher percentage (~4.7%) of their income on water compared to wealthier individuals (~0.5%). Additionally, the wealthier individuals consume over twice as much water. It is important to note that these percentages may vary across different countries (Heim, 2002). Research indicates the importance of implementing income-dependent price, full-cost price, and progressive block price systems rather than egalitarian pricing systems. On the other hand, it is worth considering that the implementation of increasing block rates, as seen in Europe, could potentially adversely affect the well-being of low-income communities in the future, especially if they are based on volume rather than per-capita measurements. In addition, the distribution of treated water and the presence of treatment plants vary among different income levels (Trenberth et al., 2014).

7.2.2.6. Storm Surge and Coastal Flooding

Storm-related hazards, such as storm surges and extratropical and tropical cyclones, are closely linked to precipitation-related hazards and pose significant risks to urban populations. Coastal regions face the threat of storm surges due to factors such as sea level rise caused by global warming, subsidence of coastal land, and intensifying storms. This poses a significant risk to the large populations residing in these areas. There have been several instances of coastal flood disasters in recent years. These include the devastating flooding caused by Super Typhoon Haiyan in the Philippines in 2013, Hurricane Sandy in New York in 2012, Cyclone Nargis in southern Myanmar in 2008, and Hurricane Katrina in New Orleans in 2005. During Hurricane Katrina, the water levels rose as high as 10 meters. Similarly, Hurricane Sandy saw water levels reaching almost 4 meters over normal tide levels (Rahmstorf, 2017).

It is important to note that individuals living in urban areas are more likely to be exposed to the dangers of storm surges and cyclones. This is primarily because urban dwellers tend to reside near or in coastal regions. Towns and cities make up a significant majority, accounting for nearly two-thirds of the population residing in coastal areas across the globe. Across all world regions, Asia has the highest percentage of its population living in the low-elevation coastal zone, with 18%. Additionally, 12% of the urban land in Asia is situated near the coast at low elevation, according to Roberts et al. (2021). In addition, it is worth noting that several major cities in Asia, including Karachi and Mumbai, are situated in coastal regions susceptible to cyclones. Coastal African cities, like Lagos and Port Harcourt in Nigeria, face the constant threat of storm surges and flooding. Several Latin American cities, including Mombasa, face similar vulnerabilities (Zheng et al., 2014).

During extreme weather events like storm surge flooding and hurricanes, vulnerable households and people are at a higher risk of being affected. This is often because these groups tend to reside in flood plains and low-lying areas, a common trend observed in many regions across the globe, as evidenced by studies conducted in various cities in Latin America, Asia, and Africa. It is important to consider the challenges faced by vulnerable households and people, including those from ethnic minorities or with low incomes, who may have limited resources to deal with any potential damage. For instance, African-American inhabitants and older people were disproportionately impacted by Hurricane Katrina. Studies indicate that there are notable differences in mortality rates between men and women during various natural and climatologic disasters. In the devastating cyclone that struck Bangladesh in 1991, the mortality rate for women was 71 per 1,000, while for men it was 15 per 1,000. There were 3,972 deaths and 2,124 deaths among men and women in Amapara, Sri Lanka, during the 2004 tsunami (Tanim and Goharian, 2021).

Inland (and coastal) flooding leads to infrastructure damages and human fatalities and poses various health risks. These include leptospirosis, diarrheal diseases, typhoid fever, cryptosporidiosis, and cholera. According to Bilskie et al. (2014), the convergence of urbanization, population growth, flooding, and climate change is expected to increase leptospirosis cases significantly. IT poses a particularly high risk for small island states, low-lying areas, and urban slums. Storms may contaminate water, including heavy metals, chemicals, and other hazardous substances, which can pose risks to communities near coastal industrial areas. In addition, a growing body of research examines the impact of extreme events, including flooding, on mental health (Hsiao et al., 2021).

7.2.3. Equity in Relation to Urban Climate Change Adaptation

Certain equity and environmental justice challenges arise when it comes to climate change in cities, such as the unequal ability to handle and adjust to these changes. It particularly impacts low-income groups, as Luo et al. (2019) highlighted. This section emphasizes inequality in connection to urban dangers brought on by maladaptation, insufficient adaptation, or failure to adjust to climate change.

The variations in the nature and scale of risks among settlements are directly related to the level of services and infrastructure available. There can be variations in risk due to insufficient or lack of services and infrastructure, which can be influenced by factors such as health, sex, and age status. Additionally, social factors, like discrimination based on gender, can also contribute to these disparities (Kondo et al., 2021). The confluence of gender roles and relations in home structures, occupations, and healthcare access caused disparities in vulnerability among low-income women, according to an analysis of the effects of the 2011 floods in Lagos. There are also disparities in risk that stem from certain groups, such as informal settlers, not having a say, and government agencies not being accountable to them It is important to consider how far adaptation measures recognize these differences, identify the most vulnerable groups, and take steps to mitigate them (Meerow and Mitchell, 2017).

Development has significantly mitigated the risk of extreme weather within most cities in high-income countries. There is widespread availability of piped treated water, sufficient drainage systems, and strict adherence to building standards to ensure structural safety. In high-income countries, cities have comprehensive storm drains and sewer systems that can handle even the most intense rainfall. Additionally, these cities have well-maintained roads, reliable healthcare systems, and efficient emergency services accessible to all residents without any disparities in their availability. These services are not solely focused on addressing climate change, but rather, they aim to enhance resilience to its impacts and bolster the financial systems and institutions that support resilience (Baron and Petersen, 2015). Within well-governed cities, significant disparities in urban services, housing quality, assets, and income between different areas can exist. However, these disparities are much less pronounced regarding the benefits provided by services and infrastructure that mitigate risks associated with climate change. These groups are typically considered vulnerable because when services and infrastructure are universally provided, it helps to minimize or eliminate their exposure to risks. Not all inequalities in risk have been adequately addressed, as evidenced by the research on environmental justice (Frantzeskaki et al., 2019).

Da Silva et al. (2012) found a strong correlation between the disproportionate effects of extreme weather events on the low-income populations in urban centers of middle- and low-income countries and the absence of risk-reducing services and infrastructure (such as emergency medical care, piped water, sanitation, efficient drains, and all-weather roads and paths). Urban governments often struggle to address service and infrastructure deficits, leading to a lack of risk-reducing infrastructure. Urban authorities in many middle- and low-income nations often face significant budget constraints and limited investment capacity. Housing development in high-risk areas such as steep slopes, riverbanks, or floodplains is worsened by inadequate building regulations and zoning and land-use practices that limit the availability of affordable housing plots. Land tenure and property rights issues are also significant factors in cities such as Kolkata, Manilla, Maputo, Dakar, Dhaka, Dar es Salaam, and Nairobi, as evidenced by documented cases (Chu, 2018).

The Fifth Assessment Report (AR5) of the IPCC emphasizes the diverse range of adaptive capacity in urban areas. On one end of the spectrum, a significant population of individuals residing in urban areas face significant challenges in adapting to climate change due to limited resources and inadequate services and infrastructure. Many people, approximately 1.5 billion, reside in urban areas that have limited capacities and face significant service and infrastructure deficits. Unfortunately, only a small fraction of the global urban population lives in cities with general provision for risk-reducing services and infrastructure and active climate change adaptation policies (Lund, 2018).

Adapting to climate change requires changing regulatory frameworks and land-use planning for zoning, infrastructure, and buildings. Implementing infrastructure investments and fiscal incentives that address projected and current climate risks is also crucial. Effective land-use management and planning are essential for addressing housing needs while safeguarding against hazardous locations and preserving ecological systems and services. It is increasingly recognized that policy interventions and adaptation processes should be gender-sensitive and consider intersectional analyses to ensure inclusivity and context-specific approaches. It is important for adaptation practices also to consider the concerns of mitigation. If these aspects are addressed, the IPCC AR5 discusses the concept of transformative adaptation (Lee and Hughes, 2017).

Insufficient government capacity is not the sole factor contributing to the lack of focus on climate change adaptation. National or city governments also make deliberate choices that play a role in this issue. The flood crisis in Thailand in 2011 clearly illustrates why policy interventions can lead to risk redistribution. City officials in Bangkok decided to divert floodwaters from the city center to other areas. However, this negatively impacted the communities outside the defenses, including the suburbs, causing a disproportionate effect (Kirshen et al., 2008). Unfortunately, the concerns of marginalized communities in urban areas are often overlooked by international agencies and national governments, resulting in a lack of attention to equity issues (Birkmann et al., 2010).

Concerning equity in the adaptation to climate change also entails preventing maladaptation in responses to climate hazards, policies, and public investments. For instance, the decisions regarding the management of floodwaters in Bangkok might be seen as a maladaptive practice due to its impact on vulnerable communities (refer to similar instances in Bangladesh and Vietnam). Several maladaptive practices can be observed, such as restricting land supplies, relocating individuals far from their workplaces, or displacing people without providing any compensation, which ultimately leads to increased housing and land expenses. These practices may lead to more people living in informal settlements and vulnerable conditions. Vulnerable groups are further impoverished due to maladaptation, often under the guise of "development." It occurs when low-income households are driven out by infrastructure projects, including highways and roads aimed at reducing deficits. Forced evictions are severe violations of human rights as they directly and indirectly infringe upon a wide range of social, political, economic, cultural, and civil rights (Yang et al., 2021). People living in hazardous areas with inadequate infrastructure often take steps to mitigate risks to their assets, homes, and households. Community involvement is crucial in addressing risks, particularly in urban centers or informal settlements with limited adaptive capacity or government interest. Klein and Juhola (2018) offer a comprehensive analysis of the various measures taken to address these challenges, including preserving local water sources, establishing sanitation facilities, and enhancing drainage systems. Implementing such measures can significantly contribute to reducing risks, particularly when implemented at a community level. However, it is important to note that these measures alone cannot address the need for a broader infrastructural network that local adaptation measures rely on, such as wastewater and water treatment plants and drainage, sewer, and water mains. However, one key takeaway from these experiences is that communities' adaptive ability and resilience may be sustainably raised by giving community-based initiatives the right assistance. It would be most effective to offer support through economic incentives and create opportunities specifically for middle- and low-income households (Olazabal et al., 2018).

7.3. EQUITY AND CLIMATE CHANGE MITIGATION

Climate change will significantly impact urban residents, encompassing risks, potential impacts, and the need for adaptation. Additionally, it is crucial to consider planning, policy, and mitigation actions to address this pressing issue. Urban mitigation actions are primarily focused in high-income countries due to their historical and ongoing contribution to greenhouse gas emissions. These cities also have the highest per capita emissions and are better equipped to take action. There is currently a lack of studies on mitigation actions in middle- and low-income countries (Shonkoff et al., 2011).

7.3.1. Spatial Planning

Regarding spatial planning, middle- and low-income countries encounter unique challenges compared to high-income countries. In many middle- and low-income countries, the policies for spatial planning struggle to keep up with the rapid population growth. Unfortunately, the city budgets are often insufficient to address the increased demand for services and infrastructure (UCLG, 2014). In addition, there is often a lack of effectiveness in spatial planning controls and limited enforcement, as (Hurlimann and March, 2012) highlighted. It is concerning that many city governments in low-income countries have not yet taken steps to address the impacts of climate change. Presently, very few cities are incorporating climate change mitigation into their strategies.

Urban revitalization is a key focus in high-income countries, where mitigation strategies aim to combat low-density development and encourage urban sprawl. One of the key principles of this planning framework is preserving open spaces, encouraging mixed land use, creating walkable neighborhoods, and promoting denser development. One of the main goals is to decrease transportation and the associated greenhouse gas emissions (Wilson, 2006).

Nevertheless, these principles have received both praise for their effectiveness in reducing traffic and criticism for their impact on housing prices, resulting in an exodus of low-income residents who cannot pay the higher rents. Specifically, implementing low-carbon construction and building standards, guided by a limited technocratic approach to addressing climate change, can hinder the advancement of socially inclusive housing policies. It could lead to housing designs that are not financially viable for middle- and low-income households, a phenomenon known as "environmental gentrification." There are concerns about the unequal allocation of domestic living space and increased exposure to air pollutants that can be associated with densification. These issues have been highlighted in studies by (Healey et al., 1999). It is important to consider these factors, especially when densification is not supported by improved public transportation. Compact city policies can have unintended consequences, potentially limiting access to open green spaces and public facilities. Research has shown that it could worsen existing inequalities. It has been observed that the availability of urban green space is often influenced by the socioeconomic and ethnic/racial makeup of a neighborhood (Foley et al., 2010). Evidently, certain communities in the United States, particularly those with low-income households, have less green space in their neighborhoods than the city averages. In South Africa, there is a significant disparity in access to public green space between low-income and black communities and predominantly white neighborhoods. Regarding urban mitigation policies centered on densification, it is crucial to prioritize a fair and balanced increase in urban density. This means ensuring that the benefits and opportunities of enhanced environments are distributed equitably (Biesbroek et al., 2009).

There is ongoing debate regarding the impact of fiscal anti-sprawl policies on residents with varying income levels. Nevertheless, it appears that development taxes, which are intended to cover infrastructure-related costs, may place a relatively lighter burden on low-income groups compared to other policies like growth and zoning management that dictate how and where development can take place (Hajer and Zonneveld, 2000).

7.3.2. Accessibility and Transport Policies

Investments in infrastructure that aim to improve private and public transportation services have proven to be an effective strategy for addressing climate change mitigation in urban areas. However, if not carefully implemented, investments can negatively affect lower-income households, such as making housing less affordable, increasing transportation costs, and reducing accessibility. Similarly, some policies focus on goals other than mitigation. For example, policies aim to reduce private car usage or enhance public transportation to alleviate congestion and reduce air pollution (Geurs and Van Wee, 2004).

Transit-oriented development (TOD) has emerged as a promising strategy to address climate change in urban areas. By enhancing public transportation accessibility, TOD has the potential to make a significant impact. Furthermore, TOD offers significant socioeconomic benefits for residents by expanding the employment catchment area and enhancing the community's overall well-being. For instance, in Medellin, the introduction of a cable car system linked isolated low-income communities with the city center, resulting in decreased violence and improved economic conditions. On the other hand, research has indicated that TOD can impact housing affordability for low-income groups. When it comes to an open housing market, it's often observed that improved access to transportation positively impacts rents and property values, especially in areas with better transportation hubs and infrastructure services. Consequently, low-income populations may be compelled to relocate to areas with worse transportation options but cheaper housing costs (Geurs et al., 2010).

One issue that needs to be addressed is the significant initial expenses associated with accessing discounted fares, such as monthly or annual transit passes, which can contribute to social inequality. Some individuals from low-income households might be required to buy daily or weekly passes, which can cost up to three times more In addition, the rise in transfer costs or the elimination of unlimited-use passes has a disproportionate impact on individuals with lower incomes, young people, and minority communities. This is because these groups tend to make transfers and take more trips more frequently than others, as highlighted in studies by (Saif et al., 2019).

Addressing crime in public transportation, harassment, and gender-based violence is crucial and requires implementing transportation-based mitigation actions that are sensitive to gender. For example, Berlin, Vienna, Malmö, Bogotá, and Montreal are cities that have successfully integrated gender considerations into their mobility plans. These cities have tried to enhance the comfort, safety, and accessibility of public transport, focusing on incorporating women's perspectives (Preston and Rajé, 2007).

Transport policies aimed at enhancing private transportation infrastructure may lead to conflicting outcomes in terms of equity. It is primarily attributed to the diverse range of policy options available, such as tax revenue recycling and other supplementary benefits. Additionally, variations in defining inequality through different methodologies further contribute to the complexity of assessing these effects. For instance, studies have found that vehicle registration fees and fuel taxes have minimal impact on income groups, although the specific design of these instruments and the measures taken to recycle the revenue can influence the results. On the other hand, pollution emission-based registration fees tend to hurt low-income drivers more than distance-based ones. This is because low-income people drive cars that pollute more than wealthy cars per mile (Vecchio et al., 2020).

There are two main pricing strategies for cordon charges: area charges, and cordon tolls. These charges are implemented either based on time or location or are designed to manage traffic flow and reduce congestion on road networks or within specific zones. The primary distinction is that the first charges a fee for a license (i.e., a particular period) to exit or enter a zone, while the second charges a fee for each crossing. Research conducted by (Vecchio et al., 2020) found that area-based schemes generally have a more positive impact on equity. This is because, with the same boundary, area-based schemes can address more demands in the network than cordon tolls (Lucas et al., 2016).

Nevertheless, implementing time-based cordon tolls in peak hours can also be a progressive approach. For instance, if revenues allotted to tolls or public transit were sufficiently low, the relative burden change in Stockholm for low-income groups was equivalent to 0.35%. According to a study by Boisjoly and El-Geneidy (2017), when considering all commuters, the burden of area-based schemes on low-income commuters decreases to 0.5%, compared to the initial burden of over 2% in Paris for low-income bikers. In general, cordon charges can bring positive results for low-income households. By improving air quality and reducing traffic congestion, these charges can contribute to environmental and social benefits and revenue recycling (Sun and Zacharias, 2020).

Speed limit schemes can result in varying burden changes, ranging from 5% to 17%. The extent of these changes depends on whether commuting patterns are taken into account in the policy design. Lastly, fees based on road usage, compared to low emission zones (LEZs), appear to be more effective in reducing environmental inequality because the latter may raise emissions nearby, which could impact populations with lower incomes. In addition, the desired improvements in air quality that LEZs aim to achieve may also be realized naturally over time as older vehicles are replaced with newer, cleaner ones (Kasraian et al., 2019). In determining who benefits from transport mitigation and spatial planning strategies, it is important to consider individual characteristics and the urban geography of policies. If the design and implementation of mitigation efforts ignore the disproportionate consequences on vulnerable groups, including low-income groups, they may worsen inequality. City-specific distributional effects necessitate a case-specific analysis at several levels, including the travel habits of various communities (Bocarejo and Oviedo, 2012).

In a broader sense, there is a concern regarding the allocation of responsibility for reducing carbon emissions in transportation systems and the fear that neo-liberal mechanisms may worsen travel patterns inequalities. Therefore, it is crucial for policymakers to proactively consider potential negative consequences and take appropriate actions to mitigate them, such as adjusting policy frameworks or implementing supplementary measures. In urban development, it has been observed that providing efficient public transport routes to all neighborhoods within a city can have multiple benefits. Not only does it help reduce private vehicle usage and overall transport volumes, but it also plays a role in addressing the welfare and distributional inequality associated with greenhouse gas emissions reduction (Pereira, 2019).

7.3.3. Renewable Energy and Waste Management

There is a lack of explicit analysis of the equity performance of sector-specific policies over urban waste management. However, certain studies provide helpful insight into specific issues like waste pickers. Waste-picking plays a crucial role in urban centers of middle- and low-income countries by significantly reducing greenhouse gas emissions. There was a study conducted in Ribeirão Pires, Brazil that demonstrated the impressive ability of the informal/cooperative recycling sector to achieve greenhouse gas emissions reductions comparable to those achieved by landfill gas capture, recycling, and formal solid waste management methods (Tan et al., 2014). Despite its sustainability, inclusivity, and integration, this survival strategy often encounters significant opposition from the authorities. When waste management and collection become a public priority, pickers are frequently displaced, and their contribution and following social impacts go unrecognized. Nevertheless, certain forward-thinking cities have developed contractual arrangements for waste-pickers. Recycling strategies in cities like Bangalore and Maputo have successfully supported waste pickers cooperatives by enhancing skills, governance, and infrastructure. These improvements have resulted in various sustainability benefits, as highlighted in a study by Eriksson et al. (2014).

It is worth noting that there may be an unequal burden shift resulting from renewable energy schemes, as low-income households tend to contribute a larger portion of their income compared to higher-income households. It highlights a significant issue with incentive-based mechanisms, as they tend to disproportionately affect low-income households who are left struggling to adapt, while high-income households can continue with their usual consumption practices with only a slight increase in cost (Bong et al., 2017).

7.4. INNOVATIONS AND LESSONS FROM IMPLEMENTATION

Many city governments have made significant strides in addressing climate change through innovative mitigation and adaptation measures. In numerous instances, cities have taken proactive measures, even without international funding and national policies, particularly in middle- and low-income nations. Recently, there has been a growing focus on adaptation. However, very few policies specifically tackle low-income group risks, particularly the billions of urban residents living in informal settlements (Bérard et al., 2015).

It is important to note that policy programs tend to distribute benefits and costs in a way that aligns with existing social class lines. It often leads to the exclusion of low-income individuals and other underrepresented groups, who may reside in informal settlements, from the process. Previous studies have highlighted this issue. This is demonstrated in disaster risk reduction, infrastructure, and housing programs. There is a clear problem with the current state of environmental policies, as they fail to adequately address the needs of certain groups and overlook the importance of "autonomous adaptation" initiatives. Therefore, these groups frequently lack support from official governance processes (Warwick et al., 1992).

Urban community actors, such as commercial enterprises and elite groups, can step in to address the gaps in formal governance systems. They may establish informal rights for the settlers and serve as intermediaries between the urban institutions and settlers. Informal governance structures can potentially turn slums into innovation hubs, fostering cohesiveness and social capital crucial for climate change adaptation. For instance, in the Dharavi section of Mumbai, individuals with expertise in environmental policy have played a crucial role in improving the slum. They have implemented waste economies, developed infrastructure, advocated for land rights, and mobilized the wider community. These measures are crucial for ensuring our collective survival and facilitating people's ability to adapt to climate change by fostering social cohesion and equitable distribution of benefits and costs. On the other hand, informality may make it harder to manage unpredictable or uncertain risks from extreme weather conditions (Collis, 1996).

In addition, informal governance can also contribute to criminal activity and result in fragmented social systems. Underrepresented and poor groups often bear the brunt of unfair treatment caused by informal institutions, which can be just as detrimental as formal governments and their structural changes. In urban areas like Dhaka, Manila, Delhi, Lagos, Nairobi, and Mumbai, the provision of essential services, including sanitation and water, is frequently dominated by commercial entities, influential individuals, or powerful community members (Wiltsey et al., 2012). Informal governance regimes have emerged as a result of the absence of formal provisions for services in informal settlements, which are often plagued by discrimination based on factors such as race, caste, class, and gender, class, caste, and race. Breaking this cycle of exploitation is the aim of numerous environmental and social justice organizations as well as policy initiatives (Akiba and Wilkinson, 2016).

It is becoming increasingly evident that city governments play a crucial role in mitigating and adapting to the effects of environmental challenges. There is a growing understanding that city governments must synchronize their disaster risk reduction, poverty reduction, and development agendas with mitigation and adaptation efforts. An analysis of the disaster risk reduction initiatives implemented by over 50 cities as part of the United Nations International Strategy for Disaster Reduction (UNISDR) "Making My City Resilient" campaign demonstrates how city governments are conceptualizing and addressing resilience to disasters, particularly concerning modifications to their community engagement and institutional frameworks. Our dedication to the cause is evident through our various initiatives, such as enforcing building codes, adapting urban planning procedures, improving informal settlements, conducting multi-hazard risk assessments, and mobilizing finance. Numerous cities prioritize including vulnerable groups, urging them to actively engage in decision-making processes, formulating policies, planning, and implementing risk reduction measures (Hagedorn et al., 2006).

These initiatives highlight the connection between addressing climate change and alleviating poverty. By implementing measures to improve informal settlements and enhance services and infrastructure, cities in middle- and low-income nations have successfully minimized the disparities in risk levels among various residents and neighborhoods. Therefore, these measures can also be considered as climate change adaptation since they contribute to developing resilience against the effects of climate change.

On the other hand, adaptation investments can neglect the needs of informal settlements and give more importance to safeguarding the formal city infrastructure. It is common for city governments in high-income and low-income countries to demolish informal settlements to create space for infrastructure projects that primarily benefit central districts and wealthier residents. The filling of natural water bodies to make space for construction or their conversion to commercial usage is an example of urban activities that go against the requirements of the most disadvantaged. These changes can significantly impact conventional drainage systems, resulting in more frequent or severe floods or waterlogging. Studies conducted in Vietnam and Bangladesh have examined the effects of climate change and potential strategies for adaptation in the regions of Hue and Satkhira, respectively. Human activity plays a significant role in waterlogging and flooding. Equity issues of utmost importance arise in situations that have far-reaching consequences for particularly vulnerable members of society. However, the contrary is also true—the impoverished might encroach upon traditional water sources by filling in riverbeds, reducing the size of water regions, and so on—increasing the risk of flooding that may impact larger urban centers (Klein and Sorra, 1996).

Many cities face challenges when implementing equity-sensitive mitigation and adaptation measures. Powerful vested interests and local political constraints often oppose these efforts, particularly if they perceive them as limiting development opportunities or impacting profits. Urban agglomerations also encounter governance challenges regarding cross-departmental and cross-municipal action. For instance, the Asian Cities Climate Change Resilience Network (ACCCRN) has encountered difficulties disseminating knowledge among various interest groups in highly politicized urban settings. Nevertheless, successful implementation requires the active support and engagement of all pertinent sectoral departments, as highlighted by (Rogers, 1995). Many cities have established a climate change focal point to facilitate the coordination of climate action among various government agencies or departments. However, placing it within the environment department may not guarantee the necessary level of attention. Environment departments often have limited influence and budgets, as seen in examples such as Sydney, Boston, and Durban.

On the other hand, New York's approach to climate change adaptation is driven by a Climate Change Adaptation Taskforce that operates under the Mayor's Long-Term Planning and Sustainability Office. The task force comprises various city and state agencies and private companies that play a crucial role in maintaining critical infrastructure. It receives guidance on Climate Change from the New York City Panel.

It is crucial to create a platform for dialogue on resilience and vulnerability within the unique context of each city. Implementing budgetary transparency and metrics to track progress on mitigation and adaptation can effectively institutionalize changes in policy and planning practice (OECD, 2012). There remains a scarcity of documentation regarding the design and execution of climate change adaptation programs and their monitoring, especially in cities in lower-middle-income and low-income countries.

However, significant attention has been given to community-based and household adaptation, highlighting its limitations and significance. Several capacity-building initiatives have been established to address these limitations, including the Urban Climate Change Research Network. Nevertheless, it is crucial to provide more specific assistance to encourage smaller cities to actively participate in climate change initiatives. Additionally, offering comprehensive training and guidance to city staff is important, specifically addressing the various equity issues prevalent in urban areas.

Municipal and city governments require assistance from multi-level governance frameworks, where national and provincial governments facilitate and back city and municipal initiatives. To direct such support, several national governments have created new regulatory frameworks, funds, and laws; many of these are aimed at lowering the risk of disaster and boosting the resilience of especially vulnerable populations.

Unfortunately, in certain countries, the existing policies, particularly at the federal level, have significant adverse effects on ethnic and low-income communities. Without the active involvement of higher-level, national, and international policy environments, urban equity policies will face significant challenges in achieving justice and addressing the underlying issues. These institutions must get involved in equity and environmental justice initiatives for meaningful progress. A local environmental justice group in the US called WE ACT focuses on addressing environmental injustices brought on by national (climate change) policy.

There is a pressing need for improved municipal, national, and international financing mechanisms to allocate resources towards local projects, focusing on assisting the most disadvantaged and vulnerable communities in adapting to climate change. Regarding this matter, the municipal government is responsible for providing financial assistance. For instance, in Cape Town, South Africa, the Kuyasa Fund offers microfinance lending for housing, focusing on assisting the most vulnerable groups, especially women. The funds are often used to enhance external house walls, enhance thermal efficiency, and expand space. In the future, there is potential to redesign this fund to promote further retrofitting, ideally in conjunction with energy-saving measures.

To ensure fairness and sustainability in climate change actions, it is crucial to focus on the continued development of social, economic, financial, regulatory, and formal institutional frameworks. Women should be acknowledged as key change agents since they solve environmental issues and reduce greenhouse gas emissions daily.

For instance, a group of 90 women heads of households in Manizales, Colombia, known as *Guardianas de la Ladera* (Guardians of the Hillside), took on traditionally male tasks to protect the environment and their homes on unstable city hillsides.

There is a project in Ikageng, a Potchefstroom township in South Africa, called Girls in Risk Reduction Leadership, which focuses on reducing the social vulnerability of underrepresented adolescent girls through practical capacity-building efforts. Experienced professionals trained girls in various areas, including disaster planning, counseling, fire safety, and public and personal health. Girls can contribute to developing risk-reduction strategies for the community, which can greatly enhance its resilience (Henstra, 2012).

7.5. EQUITY IMPACT ASSESSMENTS IN CITIES

Recent international reviews have suggested that structural equity impact assessments should be conducted to prevent any negative effects on urban equity caused by informal and formal mitigation and adaptation programs and policies. The goal is to promote sustainable and equitable climate change and environmental policy implementation, like the environmental impact assessments conducted before significant infrastructure projects in numerous nations (Caggiano et al., 2023).

There are currently four assessment approaches being discussed:

- The Equity Reference Framework (ERF) was developed by the CAN Equity Group in 2014.
- An open indicator approach.
- The template of indicators technique.
- The equity framework proposed by (Caggiano et al., 2023).

1. One proposal by the Climate Action Network is the ERF, which aims to implement the concept of common but differentiated responsibilities in global climate talks. There have been doubts about its usefulness. One important issue is the potential impact on existing inequalities or local equity when using globally and nationally aggregated indicators without considering local circumstances. Another issue that arises is the differential treatment of mitigation and adaptation. While mitigation can be substituted, adaptation is generally considered specific to each location. Equity issues have been a recurring topic of discussion at the UNFCCC Conferences from COP17 to COP19). These conferences have highlighted the importance of addressing equity concerns in adaptation and mitigation efforts. However, it is worth noting that most of the equity concerns expressed at the international level primarily revolve around different allocation strategies for climate change mitigation commitments. Nevertheless, equity issues are typically considered to be unrelated to specific locations (Laituri and Kirby, 1994).
2. The "open indicator" approach is to consider all possible indicators. Challenges with indicator-based approaches, like the ERF, underscore the differences in capabilities, impacts, and costs across different regions and the varying perspectives on what constitutes good principles and which indicators should be employed. Using an open indicator approach, each national government can choose indicators that align with its interests, potentially reducing ambition and a more favorable portrayal of its stance. It has the potential to weaken the implementation of measures. Some scholars argue in favor of using region-specific equity indicators, as (Buregeya et al., 2019) suggested.
3. A "bounded flexibility" approach allows for various commitments while adhering to established indicators. This approach creates a definitive list of indicators, established through a rigorous expert process, and grants each entity the autonomy to choose which indicators to track.

4. Karner (2016) developed an equity framework approach that analyzes how policies address equity and evaluate its baseline status. This approach has been developed through careful analysis of the regional and local policy arena and a deep understanding of values, norms, and social contexts. When planning or assessing a project or policy, it's important to consider four parameters of equity: the goals, targets, process, and content. According to Feezell et al. (2018) it is recommended that policy analysts begin their analysis by examining equity goals. Goals can vary, ranging from maximizing or enhancing equity to ensuring no harm is done or simply avoiding worsening the current situation. When establishing targets, it is crucial to consider the individuals involved and the specific social, temporal, and spatial dimensions. At this stage, making critical decisions and prioritizing a comprehensive evaluation, focusing on including vulnerable groups, is essential. Analyzing impacts on different target groups and tradeoffs between sectors and weighting or ordering conflicting rights is a crucial aspect of the equity framework. It allows for considering various factors, such as sustainability vs. growth, majority or migrant rights vs. indigenous people's rights, and future vs. current duties and responsibilities (Mabillard, 2022).

7.6. SUMMARY

This chapter explores the intricate connections between social equity and the impacts of climate change in urban settings. It examines how marginalized communities often bear the brunt of environmental degradation and climate-related disasters. By highlighting case studies and statistical data, the chapter illustrates the disparities in exposure to pollutants, access to green spaces, and resilience to extreme weather events. It also delves into policy frameworks and initiatives aimed at promoting environmental justice, advocating for equitable distribution of resources, and inclusive decision-making processes.

REFERENCES

1. Adams, C., Brown, P., Morello-Frosch, R., Brody, J. G., Rudel, R., Zota, A., & Patton, A. S. (2011). Disentangling the exposure experience: The roles of community context and report-back of environmental exposure data. *Journal of Health and Social Behavior, 52*(2), 180–196.
2. Akiba, M., & Wilkinson, B. (2016). Adopting an international innovation for teacher professional development: State and district approaches to lesson study in Florida. *Journal of Teacher Education, 67*(1), 74–93.
3. Amorim-Maia, A. T., Anguelovski, I., Chu, E., & Connolly, J. (2022). Intersectional climate justice: A conceptual pathway for bridging adaptation planning, transformative action, and social equity. *Urban Climate, 41*, 101053.
4. Baron, N., & Petersen, L. K. (2015). Understanding controversies in urban climate change adaptation: A case study of the role of homeowners in the process of climate change adaptation in Copenhagen. *Nordic Journal of Science and Technology Studies, 3*(2), 4–13.
5. Ben Selma, M., Yan, W., & Hafsi, T. (2022). Board demographic diversity, institutional context and corporate philanthropic giving. *Journal of Management and Governance, 26*(1), 99–127.
6. Bennett, N. J., Blythe, J., Tyler, S., & Ban, N. C. (2016). Communities and change in the anthropocene: Understanding social-ecological vulnerability and planning adaptations to multiple interacting exposures. *Regional Environmental Change, 16*, 907–926.
7. Bérard, E., Bonnier, C., Saulpic, O., & Zarlowski, P. (2015). Implementing managerial innovations: Lessons from two case studies. *British Journal of Healthcare Management, 21*(3), 125–129.
8. Biesbroek, G. R., Swart, R. J., & Van der Knaap, W. G. (2009). The mitigation–adaptation dichotomy and the role of spatial planning. *Habitat International, 33*(3), 230–237.
9. Bilskie, M. V., Hagen, S. C., Medeiros, S. C., & Passeri, D. L. (2014). Dynamics of sea level rise and coastal flooding on a changing landscape. *Geophysical Research Letters, 41*(3), 927–934.
10. Birkmann, J., Garschagen, M., Kraas, F., & Quang, N. (2010). Adaptive urban governance: New challenges for the second generation of urban adaptation strategies to climate change. *Sustainability Science, 5*, 185–206.
11. Bocarejo S, J. P., & Oviedo H, D. R. (2012). Transport accessibility and

social inequities: A tool for identification of mobility needs and evaluation of transport investments. *Journal of Transport Geography, 24*, 142–154.

12. Boisjoly, G., & El-Geneidy, A. M. (2017). How to get there? A critical assessment of accessibility objectives and indicators in metropolitan transportation plans. *Transport Policy, 55*, 38–50.
13. Bong, C. P. C., Ho, W. S., Hashim, H., Lim, J. S., Ho, C. S., Tan, W. S. P., & Lee, C. T. (2017). Review on the renewable energy and solid waste management policies towards biogas development in Malaysia. *Renewable and Sustainable Energy Reviews, 70*, 988–998.
14. Buregeya, J. M., Loignon, C., & Brousselle, A. (2019). Contribution to healthy places: Risks of equity free health impact assessment. *Evaluation and Program Planning, 73*, 138–145.
15. Caggiano, H., Kocakuşak, D., Kumar, P., & Tier, M. O. (2023). US cities' integration and evaluation of equity considerations into climate action plans. *npj Urban Sustainability, 3*(1), 50.
16. Cazorla, M. V., & Toman, M. A. (2010). International equity and climate change policy. In *Climate Change Economics and Policy, 7*(2), 235–247.
17. Chang, T. J., & Kleopa, X. A. (1991). A proposed method for drought monitoring. *Journal of the American Water Resources Association, 27*(2), 275–281.
18. Chapman, D. S., & Pollack, H. N. (1975). Global heat flow: A new look. *Earth and Planetary Science Letters, 28*(1), 23–32.
19. Chen, C., Doherty, M., Coffee, J., Wong, T., & Hellmann, J. (2016). Measuring the adaptation gap: A framework for evaluating climate hazards and opportunities in urban areas. *Environmental Science & Policy, 66*, 403–419.
20. Chu, E. K. (2018). Urban climate adaptation and the reshaping of state–society relations: The politics of community knowledge and mobilisation in Indore, India. *Urban Studies, 55*(8), 1766–1782.
21. Chu, E., Anguelovski, I., & Roberts, D. (2017). Climate adaptation as strategic urbanism: Assessing opportunities and uncertainties for equity and inclusive development in cities. *Cities, 60*, 378–387.
22. Collis, B. (1996). The Internet as an educational innovation: Lessons from experience with computer implementation. *Educational Technology, 36*(6), 21–30.
23. Da Silva, J., Kernaghan, S., & Luque, A. (2012). A systems approach to meeting the challenges of urban climate change. *International Journal of Urban Sustainable Development, 4*(2), 125–145.

24. Dong, C., McWilliams, J. C., Liu, Y., & Chen, D. (2014). Global heat and salt transports by eddy movement. *Nature Communications, 5*(1), 3294.

25. Eriksson, O., Bisaillon, M., Haraldsson, M., & Sundberg, J. (2014). Integrated waste management as a means to promote renewable energy. *Renewable Energy, 61*, 38–42.

26. Feezell, J. T. (2018). Agenda setting through social media: The importance of incidental news exposure and social filtering in the digital era. *Political Research Quarterly, 71*(2), 482–494.

27. Foley, M. M., Halpern, B. S., Micheli, F., Armsby, M. H., Caldwell, M. R., Crain, C. M., & Steneck, R. S. (2010). Guiding ecological principles for marine spatial planning. *Marine Policy, 34*(5), 955–966.

28. Forman, C., Muritala, I. K., Pardemann, R., & Meyer, B. (2016). Estimating the global waste heat potential. *Renewable and Sustainable Energy Reviews, 57*, 1568–1579.

29. Frantzeskaki, N., McPhearson, T., Collier, M. J., Kendal, D., Bulkeley, H., Dumitru, A., & Pintér, L. (2019). Nature-based solutions for urban climate change adaptation: Linking science, policy, and practice communities for evidence-based decision-making. *BioScience, 69*(6), 455–466.

30. Garg, K., & Nayar, S. K. (2005). When does a camera see rain? In *Tenth IEEE International Conference on Computer Vision (ICCV'05)* (Vol. 2, pp. 1067–1074).

31. Geurs, K. T., & Van Wee, B. (2004). Accessibility evaluation of land-use and transport strategies: Review and research directions. *Journal of Transport Geography, 12*(2), 127–140.

32. Geurs, K., Zondag, B., De Jong, G., & de Bok, M. (2010). Accessibility appraisal of land-use/transport policy strategies: More than just adding up travel-time savings. *Transportation Research Part D: Transport and Environment, 15*(7), 382–393.

33. Hagedorn, H., Hogan, M., Smith, J. L., Bowman, C., Curran, G. M., Espadas, D., & Sales, A. E. (2006). Lessons learned about implementing research evidence into clinical practice: Experiences from VA QUERI. *Journal of General Internal Medicine, 21*, S21–S24.

34. Haile, G. G., Tang, Q., Li, W., Liu, X., & Zhang, X. (2020). Drought: Progress in broadening its understanding. *Wiley Interdisciplinary Reviews: Water, 7*(2), e1407.

35. Hajer, M., & Zonneveld, W. (2000). Spatial planning in the network society: Rethinking the principles of planning in the Netherlands. *European Planning Studies, 8*(3), 337–355.

36. Hasterok, D., Chapman, D. S., & Davis, E. E. (2011). Oceanic heat flow: Implications for global heat loss. *Earth and Planetary Science Letters, 311*(3, 4), 386–395.

37. Healey, P., Khakee, A., Motte, A., & Needham, B. (1999). European developments in strategic spatial planning. *European Planning Studies, 7*(3), 339–355.

38. Heim Jr, R. R. (2002). A review of twentieth-century drought indices used in the United States. *Bulletin of the American Meteorological Society, 83*(8), 1149 1166.

39. Henstra, D. (2012). Toward the climate-resilient city: Extreme weather and urban climate adaptation policies in two Canadian provinces. *Journal of Comparative Policy Analysis: Research and Practice, 14*(2), 175–194.

40. Hsiao, S. C., Chiang, W. S., Jang, J. H., Wu, H. L., Lu, W. S., Chen, W. B., & Wu, Y. T. (2021). Flood risk influenced by the compound effect of storm surge and rainfall under climate change for low-lying coastal areas. *Science of the Total Environment, 764*, 144439.

41. Hurlimann, A. C., & March, A. P. (2012). The role of spatial planning in adapting to climate change. *Wiley Interdisciplinary Reviews: Climate Change, 3*(5), 477–488.

42. Ikeme, J. (2003). Equity, environmental justice, and sustainability: Incomplete approaches in climate change politics. *Global Environmental Change, 13*(3), 195–206.

43. Janin, J., & Chothia, C. (1985). Domains in proteins: Definitions, location, and structural principles. In *Methods in Enzymology* (Vol. 115, pp. 420–430).

44. Jongmans, D., & Garambois, S. (2007). Geophysical investigation of landslides: A review. *Bulletin de la Société Géologique de France, 178*(2), 101–112.

45. Kammerbauer, M., & Wamsler, C. (2017). Social inequality and marginalization in post-disaster recovery: Challenging the consensus? *International Journal of Disaster Risk Reduction, 24*, 411–418.

46. Karner, A. (2016). Planning for transportation equity in small regions: Towards meaningful performance assessment. *Transport Policy, 52*, 46–54.

47. Kasraian, D., Maat, K., & Van Wee, B. (2019). The impact of urban proximity, transport accessibility, and policy on urban growth: A longitudinal analysis over five decades. *Environment and Planning B: Urban Analytics and City Science, 46*(6), 1000–1017.

48. Keyantash, J., & Dracup, J. A. (2002). The quantification of drought: An evaluation of drought indices. *Bulletin of the American Meteorological Society, 83*(8), 1167–1180.
49. Kirshen, P., Ruth, M., & Anderson, W. (2008). Interdependencies of urban climate change impacts and adaptation strategies: A case study of Metropolitan Boston, USA. *Climatic Change, 86*(1), 105–122.
50. Klein, J., & Juhola, S. (2018). The influence of administrative traditions and governance on private involvement in urban climate change adaptation. *Review of Policy Research, 35*(6), 930–952.
51. Klein, K. J., & Sorra, J. S. (1996). The challenge of innovation implementation. *Academy of Management Review, 21*(4), 1055–1080.
52. Kogan, F. N. (1997). Global drought watch from space. *Bulletin of the American Meteorological Society, 78*(4), 621–636.
53. Kondo, K., Mabon, L., Bi, Y., Chen, Y., & Hayabuchi, Y. (2021). Balancing conflicting mitigation and adaptation behaviours of urban residents under climate change and the urban heat island effect. *Sustainable Cities and Society, 65*, 102585.
54. Korup, O., Densmore, A. L., & Schlunegger, F. (2010). The role of landslides in mountain range evolution. *Geomorphology, 120*(1, 2), 77–90.
55. Laituri, M., & Kirby, A. (1994). Finding fairness in America's cities? The search for environmental equity in everyday life. *Journal of Social Issues, 50*(3), 121–139.
56. Lee, T., & Hughes, S. (2017). Perceptions of urban climate hazards and their effects on adaptation agendas. *Mitigation and Adaptation Strategies for Global Change, 22*, 761–776.
57. Lindzen, R. S., & Farrell, B. (1980). The role of polar regions in global climate, and a new parameterization of global heat transport. *Monthly Weather Review, 108*(12), 2064–2079.
58. Liu, P., & Ma, L. (2016). Food scandals, media exposure, and citizens' safety concerns: A multilevel analysis across Chinese cities. *Food Policy, 63*, 102–111.
59. Luber, G., & McGeehin, M. (2008). Climate change and extreme heat events. *American Journal of Preventive Medicine, 35*(5), 429–435.
60. Lucas, K., Van Wee, B., & Maat, K. (2016). A method to evaluate equitable accessibility: Combining ethical theories and accessibility-based approaches. *Transportation, 43*, 473–490.
61. Lund, D. H. (2018). Governance innovations for climate change adaptation in urban Denmark. *Journal of Environmental Policy & Planning, 20*(5),

632–644.

62. Luo, X., Yu, C. W., Zhou, D., & Gu, Z. (2019). Challenges and adaptation to urban climate change in China: A viewpoint of urban climate and urban planning. *Indoor and Built Environment, 28*(9), 1157–1161.
63. Ma, L., & Christensen, T. (2019). Government trust, social trust, and citizens' risk concerns: Evidence from crisis management in China. *Public Performance & Management Review, 42*(2), 383–404.
64. Mabillard, V. (2022). Trust in government: Assessing the impact of exposure to information in a local context. *International Journal of Public Administration, 45*(9), 687–696.
65. Macdonald, A. M. (1993). Property fluxes at 30° S and their implications for the Pacific-Indian throughflow and the global heat budget. *Journal of Geophysical Research: Oceans, 98*(C4), 6851–6868.
66. Malamud, B. D., Turcotte, D. L., Guzzetti, F., & Reichenbach, P. (2004). Landslides, earthquakes, and erosion. *Earth and Planetary Science Letters, 229*(1, 2), 45–59.
67. May, C. K. (2019). Governing resilience through power: Explaining community adaptations to extreme events in coastal Louisiana. *Rural Sociology, 84*(3), 489–515.
68. McDermott, S. D., Yu, H. B., & Zurek, K. M. (2011). Turning off the lights: How dark is dark matter? *Physical Review D: Particles, Fields, Gravitation, and Cosmology, 83*(6), 063509.
69. McManus, P., Shrestha, K. K., & Yoo, D. (2014). Equity and climate change: Local adaptation issues and responses in the City of Lake Macquarie, Australia. *Urban Climate, 10*, 1–18.
70. Meerow, S., & Mitchell, C. L. (2017). Weathering the storm: The politics of urban climate change adaptation planning. *Environment and Planning A: Economy and Space, 49*(11), 2619–2627.
71. Messager, M. L., Ettinger, A. K., Murphy-Williams, M., & Levin, P. S. (2021). Fine-scale assessment of inequities in inland flood vulnerability. *Applied Geography, 133*, 102492.
72. Mukherjee, S., Mishra, A., & Trenberth, K. E. (2018). Climate change and drought: A perspective on drought indices. *Current Climate Change Reports, 4*, 145–163.
73. Olazabal, M., Chiabai, A., Foudi, S., & Neumann, M. B. (2018). Emergence of new knowledge for climate change adaptation. *Environmental Science & Policy, 83*, 46–53.
74. Pereira, R. H. (2019). Future accessibility impacts of transport policy

scenarios: Equity and sensitivity to travel time thresholds for Bus Rapid Transit expansion in Rio de Janeiro. *Journal of Transport Geography, 74*, 321–332.

75. Pietrafesa, L. J., Zhang, H., Bao, S., Gayes, P. T., & Hallstrom, J. O. (2019). Coastal flooding and inundation and inland flooding due to downstream blocking. *Journal of Marine Science and Engineering, 7*(10), 336.
76. Pollack, H. N., Hurter, S. J., & Johnson, J. R. (1993). Heat flow from the Earth's interior: Analysis of the global data set. *Reviews of Geophysics, 31*(3), 267–280.
77. Preston, J., & Rajé, F. (2007). Accessibility, mobility and transport-related social exclusion. *Journal of Transport Geography, 15*(3), 151–160.
78. Rahmstorf, S. (2017). Rising hazard of storm-surge flooding. *Proceedings of the National Academy of Sciences, 114*(45), 11806–11808.
79. Roberts, K. J., Dietrich, J. C., Wirasaet, D., Pringle, W. J., & Westerink, J. J. (2021). Dynamic load balancing for predictions of storm surge and coastal flooding. *Environmental Modelling & Software, 140*, 105045.
80. Rodríguez, C. E., Bommer, J. J., & Chandler, R. J. (1999). Earthquake-induced landslides: 1980–1997. *Soil Dynamics and Earthquake Engineering, 18*(5), 325–346.
81. Rogers, E. M. (1995). Lessons for guidelines from the diffusion of innovations. *The Joint Commission Journal on Quality Improvement, 21*(7), 324–328.
82. Saif, M. A., Zefreh, M. M., & Torok, A. (2019). Public transport accessibility: A literature review. *Periodica Polytechnica Transportation Engineering, 47*(1), 36–43.
83. Schipper, E. L. F., Ensor, J., Mukherji, A., Mirzabaev, A., Fraser, A., Harvey, B., & Shawoo, Z. (2021). Equity in climate scholarship: A manifesto for action. *Climate and Development, 13*(10), 853–856.
84. Schlosberg, D., & Collins, L. B. (2014). From environmental to climate justice: Climate change and the discourse of environmental justice. *Wiley Interdisciplinary Reviews: Climate Change, 5*(3), 359–374.
85. Shonkoff, S. B., Morello-Frosch, R., Pastor, M., & Sadd, J. (2011). The climate gap: Environmental health and equity implications of climate change and mitigation policies in California—A review of the literature. *Climatic Change, 109*, 485–503.
86. Smit, B., & Pilifosova, O. (2003). Adaptation to climate change in the context of sustainable development and equity. *Sustainable Development, 8*(9), 9.

87. Stein, C. A., & Stein, S. (1994). Constraints on hydrothermal heat flux through the oceanic lithosphere from global heat flow. *Journal of Geophysical Research: Solid Earth, 99*(B2), 3081–3095.

88. Sun, Q., Miao, C., Hanel, M., Borthwick, A. G., Duan, Q., Ji, D., & Li, H. (2019). Global heat stress on health, wildfires, and agricultural crops under different levels of climate warming. *Environment International, 128*, 125–136.

89. Sun, Z., & Zacharias, J. (2020). Transport equity as relative accessibility in a megacity: Beijing. *Transport Policy, 92*, 8–19.

90. Svoboda, M., LeComte, D., Hayes, M., Heim, R., Gleason, K., Angel, J., & Stephens, S. (2002). The drought monitor. *Bulletin of the American Meteorological Society, 83*(8), 1181–1190.

91. Tan, S. T., Hashim, H., Lim, J. S., Ho, W. S., Lee, C. T., & Yan, J. (2014). Energy and emissions benefits of renewable energy derived from municipal solid waste: Analysis of a low carbon scenario in Malaysia. *Applied Energy, 136*, 797–804.

92. Tanim, A. H., & Goharian, E. (2021). Developing a hybrid modeling and multivariate analysis framework for storm surge and runoff interactions in urban coastal flooding. *Journal of Hydrology, 595*, 125670.

93. Thieken, A. H., Samprogna Mohor, G., Kreibich, H., & Müller, M. (2022). Compound inland flood events: Different pathways, different impacts, and different coping options. *Natural Hazards and Earth System Sciences, 22*(1), 165–185.

94. Thomas, D. S., & Twyman, C. (2005). Equity and justice in climate change adaptation amongst natural-resource-dependent societies. *Global Environmental Change, 15*(2), 115–124.

95. Trenberth, K. E., & Zhang, Y. (2018). How often does it really rain? *Bulletin of the American Meteorological Society, 99*(2), 289–298.

96. Trenberth, K. E., Dai, A., Van Der Schrier, G., Jones, P. D., Barichivich, J., Briffa, K. R., & Sheffield, J. (2014). Global warming and changes in drought. *Nature Climate Change, 4*(1), 17–22.

97. Vecchio, G., Tiznado-Aitken, I., & Hurtubia, R. (2020). Transport and equity in Latin America: A critical review of socially oriented accessibility assessments. *Transport Reviews, 40*(3), 354–381.

98. Warwick, D. P., Reimers, F., & McGinn, N. (1992). The implementation of educational innovations: Lessons from Pakistan. *International Journal of Educational Development, 12*(4), 297–307.

99. Webb, P. A. U. L. (1995). The physiology of heat regulation. *American*

Journal of Physiology-Regulatory, Integrative and Comparative Physiology, 268(4), R838–R850.

100. Wiersum-Osselton, J. C., Wood, E., Bolton-Maggs, P. H. B., & Schipperus, M. R. (2014). Definitions in haemovigilance: Guiding principles and current state of development of international reference definitions. *ISBT Science Series, 9*(1), 91–97.
101. Wilhite, D. A., & Glantz, M. H. (1985). Understanding the drought phenomenon: The role of definitions. *Water International, 10*(3), 111–120.
102. Wilhite, D. A., Sivakumar, M. V., & Pulwarty, R. (2014). Managing drought risk in a changing climate: The role of national drought policy. *Weather and Climate Extremes, 3*, 4–13.
103. Wilson, E. (2006). Adapting to climate change at the local level: The spatial planning response. *Local Environment, 11*(6), 609–625.
104. Wilson, S. M., Richard, R., Joseph, L., & Williams, E. (2010). Climate change, environmental justice, and vulnerability: An exploratory spatial analysis. *Environmental Justice, 3*(1), 13–19.
105. Wiltsey Stirman, S., Kimberly, J., Cook, N., Calloway, A., Castro, F., & Charns, M. (2012). The sustainability of new programs and innovations: A review of the empirical literature and recommendations for future research. *Implementation Science, 7*, 1–19.
106. Yang, H., Lee, T., & Juhola, S. (2021). The old and the climate adaptation: Climate justice, risks, and urban adaptation plan. *Sustainable Cities and Society, 67*, 102755.
107. Zheng, F., Westra, S., Leonard, M., & Sisson, S. A. (2014). Modeling dependence between extreme rainfall and storm surge to estimate coastal flooding risk. *Water Resources Research, 50*(3), 2050–2071.

Chapter 8

INTERNATIONAL ENVIRONMENTAL LAW

Contents

INTRODUCTION

It is widely recognized that environmental laws play a crucial role in safeguarding the environment. It encompasses various legal frameworks dictating how humans interact with their environment. It covers laws controlling natural resource management, including minerals, forests, fisheries, environmental impact assessments, and environmental rules (Kim and Mackey, 2014). Environmental law focuses on safeguarding living organisms, including humans, from the potential harm caused by human activities. It encompasses the protection of ecosystems and the preservation of their essential resources and habitats (Ebbesson, 1998).

8.1. HISTORICAL DEVELOPMENT

National governments have historically enacted legislation to safeguard public health against environmental pollution. In ancient times, around AD 80, the Senate of Rome enacted legislation to safeguard the city's precious water resources, ensuring its residents had access to clean water for bathing and drinking. During the 14th century, England implemented strict regulations to address waste disposal into waterways and coal burning in London. William Penn, the Quaker leader of Pennsylvania, an English colony, in 1681, commanded that one acre of forest be kept for every five acres cleared for population. Benjamin Franklin spearheaded several efforts to reduce waste dumping during the next century. During the 19th century, the British government implemented regulations to mitigate the harmful impacts of chemical manufacturing and coal burning on the environment and public health amidst the Industrial Revolution (Michaud et al., 1998).

In the pre-20th century era, the number of international environmental agreements was limited. The agreements primarily addressed fishing rights in shared waterways, navigation, and boundary waters while neglecting ecological issues, including pollution. In the early 20th century, several conventions were established to safeguard commercial value species. These included the Convention for the Protection of Birds Useful to Agriculture (1902), which 12 European governments signed. Another significant convention was the Convention for the Preservation and Protection of Fur Seals (1911), which involved the UK, Russia, Japan, and the US (Banks, 1993).

Additionally, the Convention for the Protection of Migratory Birds (1916) was adopted by the UK (on behalf of Canada) and the US and later extended to include Mexico in 1936. During the 1930s, several countries, including the United Kingdom, Sudan, South Africa, Portugal, Italy, Egypt, and Belgium, came together to adopt the Convention for the Preservation of Flora and Fauna in their Natural State. This important agreement demonstrated their commitment to preserving Africa's natural flora and fauna by establishing national reserves and parks. France and Spain signed the convention but never followed through with ratification (Robinson and Toledo, 2012). On the other hand, Tanzania took a more proactive approach and formally adopted it in 1962. Having joined the agreement in 1939, India was bound by the document's provisions that prohibited the creation of "trophies" from any animal listed in the annex (Funkhouser, 1937).

In the 1960s, environmentalism became a crucial intellectual and political movement in the Western world. The publication of biologist Rachel Carson's Silent Spring (1962) catalyzed a profound shift in perspective in the United States. Carson's persuasive and passionate examination of chlorinated hydrocarbon pesticides and the resulting environmental damage prompted a reevaluation of various potential and actual environmental hazards. Over the years, the United States government has enacted many environmental laws. These include legislation protecting endangered species, minimizing water and air pollution, and dealing with solid-waste disposal. To ensure adherence to these laws, an Environmental Protection Agency was established. These new environmental laws significantly expanded the national government's involvement in an area previously mainly regulated by local and state authorities (Takacs, 2013).

Japan experienced a rapid reindustrialization after World War II, which unfortunately led to the indiscriminate discharge of industrial chemicals in human food chains in particular regions. In Minamata, for instance, many individuals experienced mercury poisoning from consuming fish tainted with industrial waste. The Japanese government started thinking about a comprehensive pollution-control program in the early 1960s, and in 1967, they passed the Basic Law for Environmental Pollution Control, the first comprehensive law of its kind in history. Minamata was not declared mercury-free till the close of the 20th century (Southern, 1971).

In 1971, the Convention on Wetlands of International Importance was adopted by thirty-four countries. With its origins in the city of Ramsar in Iran, the Ramsar Convention is also known as the Waterfowl Habitat Convention. The agreement, effective in 1975, currently has nearly 100 participating nations. It was mandated that all nations establish a minimum of a single protected wetland area, acknowledging the crucial contribution of wetlands in preserving ecological balance (Baranova, 1987).

After the United Nations Conference on the Human Environment in Stockholm in 1972, the UN created the UNEP as the primary international environmental organization. Despite overseeing numerous modern-day agreements, UNEP lacks the authority to enforce or impose sanctions on parties that do not comply. Despite this, the conference led to the establishment of several significant conventions. These include the Convention on International Trade in Endangered Species of Wild Fauna and Flora, and the Convention on the Prevention of Marine Pollution by Dumping of Wastes or Other Matter (Edney, 1993).

Before the Stockholm conference, Europe had been relatively sluggish in implementing legal standards for environmental protection. However, a few exceptions existed, like the UK's conservationist Countryside Act of 1968. Shortly after the UN conference in October 1972, the European Community (EC) leaders acknowledged the importance of balancing environmental protection and economic expansion. The executive branch was instrumental in shaping environmental policy throughout the following years. One notable milestone was the production of the pilot Environmental Action Programme, which set the stage for European countries to lead the way in environmental policy-making. In Germany, for instance, there was a significant shift in public opinion regarding environmental protection during the early 1980s. This change was triggered by the alarming revelation that numerous German forests had been devastated by acid rain. Since its establishment in 1980, the German Green Party has made significant strides in advocating for more stringent environmental regulations. Notably, in 1983, they achieved representation in the Bundestag, marking a significant milestone for the party. The party had entered a coalition government by the turn of the 20th century and was in charge of developing and implementing Germany's comprehensive environmental policies. Denmark, The Netherlands, and Germany, known as the "green troika," have emerged as leading pioneers in environmental law (Gassman et al., 2007).

In the 1980s, the issue of environmental pollution and its impact across borders led to negotiations and discussions on various international environmental conventions. The impact of the 1986 Chornobyl nuclear power plant accident in Ukraine (then part of the Soviet Union) was particularly noteworthy. Due to the impact of pollution, European countries in the downwind path had to implement measures to limit their populations' consumption of vegetables, meat, milk, and water. Traces of radiation have been detected in human breast milk and cow's milk in Austria. Due to the catastrophic incident, two crucial international agreements were swiftly developed in 1986. These agreements, namely the Convention on Assistance in the Case of Nuclear Accident or Radiological Emergency and the Convention on Early Notification of a Nuclear Accident, were created to guarantee prompt support and notification in the occurrence of a nuclear accident. During the next decade, a Convention on Nuclear Safety (1994) was established to encourage countries to implement fundamental guidelines for the secure functioning of land-based nuclear plants

(Rosenberg, 1970).

It is not uncommon to come across conflicting data regarding the environmental impact of human activities. The presence of scientific uncertainty can make it challenging to create and enforce environmental regulations and laws, especially during international conferences aimed at establishing universal standards. Therefore, these regulations and laws are typically crafted to be adaptable to shifts in technological advancements and scientific knowledge (Sandin, 2004). For instance, the 1985 Vienna Convention for the Protection of the Ozone Layer did not provide any specific guidelines for signatory states to follow to safeguard the environment and human health from the harmful impacts of ozone depletion. It also did not address the substances believed to be responsible for damaging the ozone layer. Similarly, the Global Warming Convention, or the Framework Convention on Climate Change, was adopted by 178 countries at the United Nations Conference on Environment and Development (popularly known as the "Earth Summit" of 1992). However, it did not establish mandatory goals for reducing the GHG emissions believed to contribute to global warming (Newig and Fritsch, 2009).

In 1995, the IPCC, established by the UNEP and World Meteorological Organization, studied changes in the Earth's temperature. Their findings pointed towards noticeable human impacts on global climate, based on the balance of evidence. Despite being hailed by environmentalists as compelling evidence of global warming, it faced criticism from certain skeptics who argued that it relied on inadequate data, exaggerated the environmental consequences of global warming, and employed unrealistic climate change models. In Japan, a conference of signatories to the UNFCCC adopted the Kyōto Protocol, which included binding emission targets for industrialized nations, marking an important milestone in global environmental policy (Sunstein, 1993). The protocol allowed developed countries to participate in emissions trading to achieve their emissions targets. The market mechanisms involved the sale of "emission reduction units" from developed countries that exceeded their emission reduction commitments to other developed countries that fell short of their targets. They could earn additional emission reduction units by financing energy-efficient projects in developing countries. It can be achieved through initiatives such as clean development mechanisms. Since its adoption, the agreement has faced strong opposition from certain countries, notably the US, that have not ratified it (Hornstein, 1992).

8.2. LEVELS OF ENVIRONMENTAL LAW

Environmental law encompasses various levels of governance and is not solely reliant on treaties, conventions, and international declarations. Most environmental law consists of statutes, laws created by legislative bodies, and regulations, which are rules made by government agencies to preserve the environment (French, 2000).

Furthermore, numerous countries have incorporated provisions for environmental quality within their national constitutions. For instance, since 1994, environmental protection has become a fundamental aspect of the German Grundgesetz ("Basic Law"). It explicitly mandates that the government safeguard the natural foundations of life for future generations. Likewise, the Chinese constitution emphasizes the importance of responsible natural resource utilization and endangered plants and wildlife preservation. The South African constitution acknowledges the fundamental right to a healthy and safe environment for future and current generations. Similarly, the Bulgarian constitution guarantees the right to a favorable and healthy environment that adheres to established regulations and standards. Lastly, the Chilean constitution enshrines the right to reside in an environment free from contamination (Palmer, 1992).

Environmental law is also reflected in the rulings of courts at the local, national, and international levels. Arbitrated decisions, like the Trail Smelter arbitration (1941), have addressed the operation of a smelter near the international border between the U.S. state of Washington, Canada, and British Columbia. This particular case established the principle that no state can use its territory in a way that causes harm to another state or its persons and properties. Environmental law can also be found in the rulings of national courts. In the case of the Scenic Hudson Preservation Conference v. Federal Power Commission (1965), the U.S. federal appeals court voided a Federal Power Commission license to construct a pumped-storage hydroelectric plant. This plant, which would have caused significant environmental damage, was planned to be built in an area of exceptional natural beauty. It serves as an example of how decisions made by federal agencies can be effectively challenged in the courts. One notable local case was the National Audubon Society v. Superior Court (1976), where the California Supreme Court imposed strict limitations on Los Angeles' water diversion practices, specifically regarding Mono Lake in the eastern desert of California (Brunnée, 2004).

8.3. TYPES OF ENVIRONMENTAL LAW

8.3.1. Command-and-Control Legislation

Environmental law is typically classified under the "command and control" laws. Environmental laws usually consist of three components: identifying environmentally harmful activities, setting specific standards or conditions, and prohibiting any activity that doesn't meet the required standards or conditions. One example is the US Federal Water Pollution Control Act (1972), which regulates the release of pollutants into the country's navigable waters. The definitions of all three terms are clearly outlined in the agency and statute regulations, collectively identifying the specific activities subject to environmental regulation. In 1983, Germany implemented a national emission-control law that established clear air emission thresholds based on the type and age of power plants. Most environmental laws strictly forbid activities that fail to meet the specified standards or conditions. Many consider it illegal to violate such standards with "knowing" intent intentionally (Aalders and Wilthagen, 1997).

One can easily identify regulated activities by looking at instances where pollutants are released into the environment, such as groundwater, water, and air pollution. Additionally, regulations are in place to oversee activities that carry a substantial potential for releasing harmful pollutants. These include logging, the sale of pesticides, and the transportation of hazardous waste. Environmental laws typically set particular thresholds for allowable pollution in cases of actual discharges. In situations with a risk of discharge, these laws often determine management practices to minimize that risk (Blackman and Liu, 2018).

There are typically two types of standards that are applied to actual discharges. The first type is environmental quality standards, which establish the maximum allowable amount of regulated pollutants in the water or air. The second type is emission standards, which control the amount of pollutants any source can discharge into the environment. Comprehensive environmental laws typically incorporate discharge standards and environmental quality, aiming to align their application to accomplish a specific environmental quality objective. There are two types of environmental-quality goals: narrative and numerical. Setting numerical targets involves establishing a precise limit on the amount of a pollutant considered acceptable within a given timeframe. For example, a target may specify that no more than 10 micrograms of CO can be present per cubic meter of air over eight hours. It is important to ensure that the receiving water or air is suitable for its intended use, such as swimming, according to narrative standards (Kunreuther et al., 2002).

The management practices recommended for activities that pose a risk of discharge are context-specific and diverse. For instance, the 1991 US Resource Conservation and Recovery Act (1991) necessitates drip pads for containers that accumulate or store hazardous waste. Similarly, the 1990 US Oil Pollution Act (1990) requires all oil tankers of a specific age and size operating in U.S. waters to be double-hulled (Job et al., 2007).

One area that is subject to regulation under command-and-control legislation is environmentally detrimental trade activities. One of the most well-established regulations pertains to wildlife trade. One example is the 1973 Convention on International Trade in Endangered Species of Wild Fauna and Flora (CITES, 1973), which grants signatories the authority to identify species at risk of extinction and may be impacted by trade. After an animal or plant species is classified as endangered, countries typically must restrict the export or import of that species unless there are specific limited circumstances (Batty and Hilton, 2003). In 1989, the African elephant was listed as a protected species, significantly restricting the African ivory trade. This addition ultimately resulted in EC and Kenya implementing a complete ban on such trade. At this point, the US banned the African ivory trade and recognized them as endangered in the 1978 Federal Endangered Species Act. Despite implementing various measures, certain countries, like Japan, did not enforce a ban on ivory imports or declined to prohibit ivory exports (such as Zimbabwe, South Africa, Namibia, and Botswana). Consequently, elephants remained vulnerable to threats posed by smugglers and poachers (Li et al., 2019).

8.3.2. Environmental Assessment Mandates

Environmental assessment mandates are a crucial aspect of environmental law. These mandates typically serve three purposes:

1. Determining when an action has a significant enough environmental impact to require an assessment,
2. Setting particular targets for the mandated assessment,
3. Establishing requirements to guarantee that the assessment is taken into account when deciding whether to pursue an alternative or to continue with the original action.

Mandated environmental assessments work differently than command-and-control regulations that directly restrict environmental discharges (Stinchcombe and Gibson, 2001). They enhance the quality and quantity of publicly accessible information about the environmental impacts of planned actions, indirectly safeguarding the environment. This information has the potential to foster greater public engagement in environmental policy development and enhance governments' decision-making process (Webler and Tuler, 2006).

It is mandated by the 1969 United States National Environmental Policy Act that an environmental impact statement must be prepared for any "major federal action that has a significant impact on the human environment quality." It is crucial to thoroughly evaluate the environmental consequences of the suggested action and explore various alternatives, including the option of taking no action. Under the 1969 act, federal agencies must adhere to the Council on Environmental Quality regulations. These regulations state that agencies must wait until the environmental impact statements are completed before deciding to eliminate alternative options (Berke et al., 1997). Similarly, the EU mandates an environmental impact assessment in two specific project categories. Projects classified as "annex-I Projects," such as thermal power stations, toxic waste landfills, and oil refineries with heat output of at least 300 megawatts, are typically required to meet certain environmental standards. On the other hand, "annex-II Projects," which include activities in paper, wood, leather, textile, food, and chemical industries, are only subject to an environmental impact assessment if member states deem it necessary based on their specific characteristics. The project's indirect and direct effects on landscape, climate, air, water, soil, flora, fauna, and humans, and the interactions between them must be described and evaluated in these evaluations (Eckert et al., 2020).

8.3.3. Economic Incentives

Economic instruments are widely used to incentivize environmental protection in environmental law. Various incentives can be implemented to reduce pollution and promote environmental protection. These include subsidies for clean practices and technologies, pollution taxes, and the establishment of markets focused on pollution control or environmental protection. Several countries, including Sweden, The Netherlands, and Denmark, have implemented taxes on CO_2 emissions. The European Union has also discussed the possibility of implementing a similar tax at a supranational level to address the climate change issue (Town et al., 2005). Water pollution laws enacted in 1972 in the United States gave local governments financial support to modernize governmental sewage treatment facilities. In 1980, the U.S. government responded to the growing national concern over industrial pollution, particularly highlighted by the Love Canal incident in Niagara Falls, New York. They established a federal "superfund" to address the issue, utilizing funds from general revenues, general corporate income, crude oil, and petrochemical feedstock taxes. This initiative successfully funded the cleanup of over 1,000 sites contaminated by hazardous substances (Hahn and Stavins, 1992).

In the 1990s, the United States saw the implementation of "tradable allowance schemes" that allowed companies to trade "pollution credits" – legal rights for generating specific amounts of pollution. One of the most complex and comprehensive programs in this field was established under the 1990 Clean Air Act. Its main objective was to decrease SO_2 emissions from power plants that burn fossil fuels. According to supporters, the program would offer financial incentives to cleaner plants, enabling them to sell their surplus credits on the market. This incentive would allow less environmentally friendly plants to continue operating while transitioning to cleaner technologies (Rode et al., 2015).

8.3.4. Set-Aside Schemes

Another effective approach to safeguarding the environment involves preserving waters and lands pristinely. In the United States, for instance, a significant portion (one-third) of the federally owned land requires the approval of a federal agency for development. Europe boasts a vast network of preserves and national parks encompassing private and public land. Similarly, eastern and southern Africa have many extensive national parks that protect wildlife. One could argue that the extensive set of regulations governing the utilization of public resources and lands owned by the public falls under the "environmental law." However, some argue against it (Macdonald et al., 2007).

Various areas of law can be described as regulatory and "set aside." For instance, global initiatives to protect wetlands have prioritized the designation of ecologically significant areas, such as wetlands, and implementing regulations to govern their utilization. According to the Ramsar Convention, wetlands are recognized as valuable resources with recreational, scientific, cultural, and economic importance. In the US, the Clean Water Act includes a provision prohibiting unpermitted discharges of "fill and dredge material" into any water bodies within the country (Firbank et al., 2003).

8.4. ENVIRONMENTAL LAW PRINCIPLES

Modern environmental law has been greatly influenced by key publications like the 1992 Rio Declaration from the Earth Summit and the 1987 Our Common Future by the World Commission on Environment and Development. These documents have played a significant role in shaping the concepts and principles that guide the application and design of environmental policies today (Palmer, 1992).

8.4.1. The Precautionary Principle

As mentioned, environmental law often deals with complex issues due to significant scientific uncertainty. Regarding various activities that involve altering the environment, it can be quite challenging to accurately predict the impact on human health or the environment's quality. For instance, it is typically impossible to predict whether the development of oil in an environmentally sensitive area will seriously affect the local wildlife, whether a given level of water pollution will decrease the population of healthy fish, or whether a given level of air pollution will raise the mortality rates from respiratory illnesses (Kriebel et al., 2001). It is important to prioritize taking action to control activities that may have environmentally harmful effects, even if there is only a strong suspicion of their impact. Waiting for indisputable scientific evidence may not be the most effective approach according to the precautionary principle. This idea is stated in the Rio Declaration, which states that cost-effective steps to prevent environmental degradation will not be delayed in cases where there are threats of irreversible or serious damage, and a lack of absolute scientific certainty is not acceptable as a reason (Peter and Kaszuba, 2002). In the US, habitat-conservation plans necessary in the Endangered Species Act were designed to incorporate the precautionary principle. In 1989, the EC made a significant move by implementing the precautionary principle prohibiting U.S. hormone-fed beef import. This decision was later reinforced in 2000 when the organization officially recognized the principle as a fundamental aspect of international law. In 1999, New Zealand and Australia took a stand against Japan, accusing them of overfishing southern Bluefin tuna. The precautionary principle was invoked to support their case (Majone, 2002).

8.4.2. The Prevention Principle

Preventing environmental harm is an environmentally friendly, easier, and cheaper approach compared to reacting to existing environmental damage. It is particularly important given the tendency to draft environmental legislation in response to emergencies. The prevention principle forms the basis of laws that govern the various aspects of pesticide use and hazardous waste management. This idea was the cornerstone of the 1989 Basel Convention on the Control of Transboundary Movements of Hazardous Wastes and their Disposal. It aimed to minimize hazardous waste generation and prevent unlawful dumping. In 1983, the EC's Third Environmental Action Program adopted the prevention principle as a crucial component (Westin and Nilstun, 2006).

8.4.3. The "Polluter Pays" Principle

The "polluter pays" theory has dominated thinking in environmental law since the early 1970s. According to numerous economists, a significant amount of environmental damage is attributed to producers who shift the expenses of their activities onto others. For instance, factories that dump untreated chemicals into a river or release unfiltered exhaust into the atmosphere often pay minimal fees for waste disposal. On the contrary, the entire community bears the burden of the cost of waste disposal in the shape of pollution (Woerdman et al., 2008). Just like a driver is responsible for paying for maintenance and fuel, they externalize the costs of the harmful gases emitted from their vehicle's tailpipe to others.

Therefore, numerous environmental regulations aim to ensure that polluters are held accountable for the actual expenses associated with their pollution, even though determining these costs can often be challenging. These measures incentivize polluters to adopt cleaner technologies and production methods. U.S. laws mandate the cleanup of hazardous substance releases, including oil, based on the principle that the responsible party should bear the cost. One such regulation, the Oil Pollution Act (1990), was enacted in response to the 1989 oil leak in Alaska's Prince William Sound, which involved around 41 million liters (11 million gallons) (Huber and Wirl, 1998). The principle of "polluter pays" is a guiding force behind the policies of numerous governments worldwide, including the EU. In Germany, a 1991 ordinance placed the burden of disposing or recycling product packaging on businesses until the end of the product's life cycle. However, this regulation was deemed unconstitutional by the German Federal Constitutional Court. Environmental policies have also been implemented at the state or regional level. For instance, in 1996, Florida incorporated a "polluter pays" provision into its constitution to safeguard the environmentally sensitive Everglades (Barthakur, 2021).

8.4.4. The Integration Principle

It is crucial to carefully consider the likely consequences of decisions that impact the environment to ensure effective environmental protection. Many jurisdictions, including the EU and the US, and business organizations like the U.S. Chamber of Commerce have incorporated environmental factors into their decision-making processes. It has been achieved through the implementation of environmental provisions like environmental-impact-assessment mandates (Siegler and Chen, 2008).

8.4.5. The Public Participation Principle

Public input is typically incorporated into decisions regarding environmental protection. Government decisions regarding environmental standards, permits for environmentally damaging operations, or the preservation of resources are made after the decision has been publicly and formally announced. The public can then participate in hearings or provide written comments to influence the decision. It is common for citizens in various countries to have the ability to legally contest government decisions that impact the environment, either through administrative channels or court proceedings. Such citizen lawsuits have emerged as a crucial element in environmental decision-making, both internationally and nationally (Innes and Booher, 2004).

Laws requiring broad public access to government environmental data have made it easier for the public to participate in environmental decision-making in

North America and Europe. Various international-level measures include the 1998 Århus Convention and the Rio Declaration. These agreements aimed to increase the availability of environmental information to the public and improve public participation in government decisions related to the environment. In the 1990s, the Internet emerged as a major platform for sharing environmental information with the public (Hourdequin et al., 2012).

8.5. SUSTAINABLE DEVELOPMENT

Sustainable development as a means of economic planning promotes economic expansion while maintaining environmental quality for generations to come. Although sustainable development gained significant popularity in the late 20th century, its practical application faced challenges. The effectiveness of long-term sustainability analyses varied depending on the specific resources under consideration. Take, for instance, the case of a forest that can continuously produce timber but fails to support native bird populations. Similarly, a mineral deposit that eventually runs out can still sustain communities to some extent. The 1992 Earth Summit strongly emphasized sustainability, which later became a key focus in numerous environmental studies (Parris and Kates, 2003).

Ecotourism is a crucial aspect of sustainable development law. While pollution and misuse of natural resources threaten the environment, tourism may also raise awareness of fragile and unique ecosystems worldwide and provide financial incentives for environmental protection in developing nations. A charter was adopted during the 1995 World Conference on Sustainable Tourism, which took place in the Canary Islands on the island of Lanzarote. This charter aimed to promote the development of laws that would simultaneously safeguard the environment and foster economic growth through tourism. After two years, the Malé Declaration on Sustainable Tourism was established by 27 Asian-Pacific countries. This declaration aimed to promote environmental ethics in tourism, minimize waste, preserve cultural and natural diversity, and encourage local community engagement and growth. The World Tourism Organization designated 2002 as the International Year of Ecotourism to draw attention to the growing significance of sustainable tourism (Omer, 2008).

8.6. CURRENT TRENDS AND PROSPECTS

Despite the conclusion of numerous international environmental treaties, it remains challenging to achieve effective agreements for various reasons. Given the global nature of environmental problems, multiple governments must collaborate to address these issues effectively. However, it is not uncommon for these governments to have significant disagreements on key aspects of environmental policy. In addition, due to the potential economic and social hardships that often accompany the implementation of environmental measures, numerous countries, especially those in the developing world, have shown hesitancy in joining environmental treaties. Over the years, many environmental treaties have included provisions to promote their adoption by developing nations. These measures encompass varying implementation obligations and schedules, technology transfer, and financial cooperation (Yu and Chen, 2016).

Compliance poses a significant obstacle to the efficacy of environmental treaties. While treaties can aim to enforce compliance through measures like sanctions, their effectiveness is often limited. This is because countries already in compliance with a treaty might be unable or unwilling to impose the sanctions outlined in the agreement. Typically, most countries prioritize maintaining their reputation in the international community over the potential consequences of sanctions when violating their international obligations. Establishing enforcement mechanisms beyond sanctions has proven challenging due to the reluctance of countries to relinquish elements of their national sovereignty to international or foreign organizations (Chernysh et al., 2021). In most agreements, enforcement is typically regarded as a domestic matter, giving each country the freedom to determine compliance based on its national interest. Given the challenges at hand, it is evident that the significance of international environmental agreements and treaties will continue to rise as global environmental issues become increasingly urgent (Herrera, 2008).

There are still many areas within international environmental law that require further development. International agreements have played a role in bringing consistency to the regulations and laws surrounding environmentally harmful activities across different countries. However, it's important to note that significant differences exist in the regulations that apply to certain activities. Due to the cross-border nature of environmentally harmful activities, inconsistent laws have created situations where certain legal activities in some countries lead to unacceptably high levels of environmental damage in neighboring countries (Guyomard et al., 2013).

This issue became especially concerning with the implementation of free trade agreements starting in the early 1990s. For instance, NAFTA led to the establishment of numerous maquiladoras. These factories are owned by Mexican and U.S. corporations and are located in Mexico within a free trade zone along the U.S. Mexican border, spanning 100 km (60 miles). Due to the lack of political determination and inadequate resources from Mexico's government, the maquiladoras could contaminate the surrounding areas without facing consequences (Arbuthnot, 2008). They frequently disposed of hazardous wastes directly into waterways or on the ground, which then flowed into U.S.

territory. Before adopting NAFTA in 1992, negotiators recognized the potential issues and included a "side agreement" to the treaty—this agreement aimed to promote environmental cooperation among the signatory states. Meanwhile, there were growing concerns about the link between environmental degradation and free trade agreements in Europe. These concerns fueled opposition to the Maastricht Treaty, which established the EU and expanded its jurisdiction (Tscharntke et al., 2011).

8.7. SUMMARY

This chapter delves into the global framework governing the protection of the environment through legal measures. It explores the origins and evolution of international environmental agreements and treaties, highlighting their role in addressing transboundary environmental issues such as climate change, biodiversity loss, and pollution. The chapter examines key international conventions. It also discusses the mechanisms for enforcement and compliance, the challenges in achieving global cooperation, and the influence of international organizations like the United Nations. Additionally, the chapter analyzes the interplay between international environmental law and national legal systems, emphasizing the importance of integrating international principles into domestic policies to achieve sustainable development and environmental justice.

REFERENCES

1. Aalders, M., & Wilthagen, T. (1997). Moving beyond command and control: Reflexivity in the regulation of occupational safety and health and the environment. *Law & Policy, 19*(4), 415–443.
2. Arbuthnot, K. (2008). A command gap? A practitioner's analysis of the value of comparisons between the UK's military and emergency services' command and control models in the context of UK resilience operations. *Journal of Contingencies and Crisis Management, 16*(4), 186–194.
3. Banks, J. A. (1993). Multicultural education: Historical development, dimensions, and practice. *Review of Research in Education, 19*, 3–49.
4. Baranova, M. A. (1987). Historical development of the present classification of morphological types of stomates. *The Botanical Review, 53*, 53–79.
5. Barthakur, A. (2021). Polluter pays principle as the key element to environmental law. *International Journal of Scientific and Research Publications, 11*(3), 274–277.
6. Batty, C., & Hilton, J. (2003). From command and control to self-confidence in government: Meeting the challenge in Doncaster. *International Review of Administrative Sciences, 69*(2), 161–172.
7. Berke, P. R., Dixon, J., & Ericksen, N. (1997). Coercive and cooperative intergovernmental mandates: A comparative analysis of Florida and New Zealand environmental plans. *Environment and Planning B: Planning and Design, 24*(3), 451–468.
8. Blackman, A., Li, Z., & Liu, A. A. (2018). Efficacy of command-and-control and market-based environmental regulation in developing countries. *Annual Review of Resource Economics, 10*(1), 381–404.
9. Brunnée, J. (2004). The United States and international environmental law: Living with an elephant. *European Journal of International Law, 15*(4), 617–649.
10. Chernysh, Y., Yakhnenko, O., Chubur, V., & Roubík, H. (2021). Phosphogypsum recycling: A review of environmental issues, current trends, and prospects. *Applied Sciences, 11*(4), 1575.
11. Ebbesson, J. (1998). The notion of public participation in international environmental law. *Yearbook of International Environmental Law, 8*(1), 51.
12. Eckert, L. E., Claxton, N. X., Owens, C., Johnston, A., Ban, N. C., Moola, F., & Darimont, C. T. (2020). Indigenous knowledge and federal environmental assessments in Canada: Applying past lessons to the 2019 impact assessment act. *Facets, 5*(1), 67–90.

13. Edney, M. H. (1993). Cartography without progress: Reinterpreting the nature and historical development of mapmaking. *Cartographica: The International Journal for Geographic Information and Geovisualization, 30*(2, 3), 54–68.

14. Firbank, L. G., Smart, S. M., Crabb, J., Critchley, C. N. R., Fowbert, J. W., Fuller, R. J., & Hill, M. O. (2003). Agronomic and ecological costs and benefits of set-aside in England. *Agriculture, Ecosystems & Environment, 95*(1), 73–85.

15. French, D. (2000). Developing states and international environmental law: The importance of differentiated responsibilities. *International & Comparative Law Quarterly, 49*(1), 35–60.

16. Funkhouser, H. G. (1937). Historical development of the graphical representation of statistical data. *Osiris, 3*, 269–404.

17. Gassman, P. W., Reyes, M. R., Green, C. H., & Arnold, J. G. (2007). The soil and water assessment tool: Historical development, applications, and future research directions. *Transactions of the ASABE, 50*(4), 1211–1250.

18. Guyomard, H., Manceron, S., & Peyraud, J. L. (2013). Trade in feed grains, animals, and animal products: Current trends, future prospects, and main issues. *Animal Frontiers, 3*(1), 14–18.

19. Hahn, R. W., & Stavins, R. N. (1992). Economic incentives for environmental protection: Integrating theory and practice. *The American Economic Review, 82*(2), 464–468.

20. Herrera, F. (2008). Genetic fuzzy systems: Taxonomy, current research trends, and prospects. *Evolutionary Intelligence, 1*, 27–46.

21. Hornstein, D. T. (1992). Reclaiming environmental law: A normative critique of comparative risk analysis. *Columbia Law Review, 92*(3), 562–633.

22. Hourdequin, M., Landres, P., Hanson, M. J., & Craig, D. R. (2012). Ethical implications of democratic theory for US public participation in environmental impact assessment. *Environmental Impact Assessment Review, 35*, 37–44.

23. Huber, C., & Wirl, F. (1998). The polluter pays versus the pollutee pays principle under asymmetric information. *Journal of Environmental Economics and Management, 35*(1), 69–87.

24. Innes, J. E., & Booher, D. E. (2004). Reframing public participation: Strategies for the 21st century. *Planning Theory & Practice, 5*(4), 419–436.

25. Job, J., Stout, A., & Smith, R. (2007). Culture change in three taxation

administrations: From command and control to responsive regulation. *Law & Policy, 29*(1), 84–101.

26. Kim, R. E., & Mackey, B. (2014). International environmental law as a complex adaptive system. *International Environmental Agreements: Politics, Law and Economics, 14*, 5–24.
27. Kriebel, D., Tickner, J., Epstein, P., Lemons, J., Levins, R., Loechler, E. L., & Stoto, M. (2001). The precautionary principle in environmental science. *Environmental Health Perspectives, 109*(9), 871–876.
28. Kunreuther, H. C., McNulty, P. J., & Kang, Y. (2002). Third-party inspection as an alternative to command and control regulation. *Risk Analysis, 22*(2), 309–318.
29. Li, W., Gu, Y., Liu, F., & Li, C. (2019). The effect of command-and-control regulation on environmental technological innovation in China: A spatial econometric approach. *Environmental Science and Pollution Research, 26*, 34789–34800.
30. Macdonald, D. W., Tattersall, F. H., Service, K. M., Firbank, L. G., & Feber, R. E. (2007). Mammals, agri-environment schemes and set-aside: What are the putative benefits? *Mammal Review, 37*(4), 259–277.
31. Majone, G. (2002). The precautionary principle and its policy implications. *JCMS: Journal of Common Market Studies, 40*(1), 89–109.
32. Michaud, R., Lehman, W. F., & Rumbaugh, M. (1988). World distribution and historical development. *Alfalfa and Alfalfa Improvement, 29*, 25–91.
33. Newig, J., & Fritsch, O. (2009). Environmental governance: Participatory, multi-level – and effective? *Environmental Policy and Governance, 19*(3), 197–214.
34. Omer, A. M. (2008). Energy, environment and sustainable development. *Renewable and Sustainable Energy Reviews, 12*(9), 2265–2300.
35. Palmer, G. (1992). New ways to make international environmental law. *American Journal of International Law, 86*(2), 259–283.
36. Parris, T. M., & Kates, R. W. (2003). Characterizing and measuring sustainable development. *Annual Review of Environment and Resources, 28*(1), 559–586.
37. Peter, L. D., & Kaszuba, M. (2002). Implementing the precautionary principle. *Science of the Total Environment, 288*(1, 2), 155–165.
38. Robinson, D. H., & Toledo, A. H. (2012). Historical development of modern anesthesia. *Journal of Investi. Surgery.*
39. Rode, J., Gómez-Baggethun, E., & Krause, T. (2015). Motivation crowding

by economic incentives in conservation policy: A review of the empirical evidence. *Ecological Economics, 117*, 270–282.

40. Rosenberg, N. (1970). Economic development and the transfer of technology: Some historical perspectives. *Technology and Culture, 11*(4), 550–575.
41. Sandin, P. (2004). The precautionary principle and the concept of precaution. *Environmental Values, 13*(4), 461–475.
42. Siegler, R. S., & Chen, Z. (2008). Differentiation and integration: Guiding principles for analyzing cognitive change. *Developmental Science, 11*(4), 433–448.
43. Southern, R. W. (1971). Presidential address: Aspects of the European tradition of historical writing: 2. Hugh of St Victor and the idea of historical development. *Transactions of the Royal Historical Society, 21*, 159–179.
44. Stinchcombe, K., & Gibson, R. B. (2001). Strategic environmental assessment as a means of pursuing sustainability: Ten advantages and ten challenges. *Journal of Environmental Assessment Policy and Management, 3*(03), 343–372.
45. Sunstein, C. R. (1993). Endogenous preferences, environmental law. *The Journal of Legal Studies, 22*(2), 217–254.
46. Takacs, L. (2013). The historical development of mechanochemistry. *Chemical Society Reviews, 42*(18), 7649–7659.
47. Town, R., Kane, R., Johnson, P., & Butler, M. (2005). Economic incentives and physicians' delivery of preventive care: A systematic review. *American Journal of Preventive Medicine, 28*(2), 234–240.
48. Tscharntke, T., Batáry, P., & Dormann, C. F. (2011). Set-aside management: How do succession, sowing patterns, and landscape context affect biodiversity? *Agriculture, Ecosystems & Environment, 143*(1), 37–44.
49. Webler, T., & Tuler, S. (2006). Four perspectives on public participation process in environmental assessment and decision making: Combined results from 10 case studies. *Policy Studies Journal, 34*(4), 699–722.
50. Westin, L., & Nilstun, T. (2006). Principles help to analyze but often give no solution—Secondary prevention after a cardiac event. *Health Care Analysis, 14*, 111–117.
51. Woerdman, E., Arcuri, A., & Clò, S. (2008). Emissions trading and the polluter-pays principle: Do polluters pay under grandfathering? *Review of Law & Economics, 4*(2), 565–590.
52. Yu, S., & Chen, P. Y. (2016). Emerging memory technologies: Recent trends and prospects. *IEEE Solid-State Circuits Magazine, 8*(2), 43–56.

INDEX

D

E

Kolibri 360 GmbH
Ferdinand-Jühlke-Straße 7
99095 Erfurt, Deutschland
produktsicherheit@kolibri360.de